Introduction to Analog and Digital Communication

RIVER PUBLISHERS SERIES IN COMMUNICATIONS
Volume 46

Series Editors

ABBAS JAMALIPOUR
The University of Sydney
Australia

MARINA RUGGIERI
University of Rome Tor Vergata
Italy

HOMAYOUN NIKOOKAR
Delft University of Technology
The Netherlands

The "River Publishers Series in Communications" is a series of comprehensive academic and professional books which focus on communication and network systems. The series focuses on topics ranging from the theory and use of systems involving all terminals, computers, and information processors; wired and wireless networks; and network layouts, protocols, architectures, and implementations. Furthermore, developments toward new market demands in systems, products, and technologies such as personal communications services, multimedia systems, enterprise networks, and optical communications systems are also covered.

Books published in the series include research monographs, edited volumes, handbooks and textbooks. The books provide professionals, researchers, educators, and advanced students in the field with an invaluable insight into the latest research and developments.

Topics covered in the series include, but are by no means restricted to the following:

- Wireless Communications
- Networks
- Security
- Antennas & Propagation
- Microwaves
- Software Defined Radio

For a list of other books in this series, visit www.riverpublishers.com
http://riverpublishers.com/series.php?msg=Communications

Introduction to Analog and Digital Communication

M. A. Bhagyaveni

Anna University
College of Engineering, Guindy, Chennai, India

R. Kalidoss

SSN College of Engineering
Chennai, India

K. S. Vishvaksenan

SSN College of Engineering
Chennai, India

River Publishers

Published, sold and distributed by:
River Publishers
Niels Jernes Vej 10
9220 Aalborg Ø
Denmark

River Publishers
Lange Geer 44
2611 PW Delft
The Netherlands

Tel.: +45369953197
www.riverpublishers.com

ISBN: 978-87-93379-33-6 (Hardback)
        978-87-93379-32-9 (Ebook)

©2016 River Publishers

Contents

PART III: Pulse and Data Communication

PART V: Multi-User Radio Communication

Preface

One of the main consequences of the unprecedented growth of analog and digital communications is the drastic increase in the number of professions that demands an in-depth knowledge of these concepts. Thus, this gives rise to the augmentation in the number and types of students opting for such courses. The diverse nature of the educational background of students interested to pursue courses on communication calls for a textbook that enables even neophytes to understand the subject. On the other hand, it must also pose challenges to the expert readers.

This book is designed to serve as a textbook or a reference material to second-year engineering undergraduates in computer science, electrical, telecommunication, and information technology. The needs of involved students interested in design of analog and digital systems are also catered to in this book.

Since this book is primarily aimed at students who have little experience in the field of communication, we use the bottom-top approach. The basics of analog communication are laid out before proceeding towards digital communication systems. The numerous examples enlisted in this book serve to provide an intuitive understanding of the theory under discussion.

Modulation and Demodulation of analog signals are explained in Chapters 1 and 2. Chapter 1 deals with Amplitude Modulation (AM), while Chapter 2 focuses on Frequency Modulation (FM) and Phase Modulation (PM).

Digital signal transmission via carrier modulation is described in Chapter 3. Amplitude Shift Keying (ASK), Phase Shift Keying (PSK), Frequency Shift Keying (FSK), Quadrature Phase Shift Keying (QPSK), Quadrature Amplitude Modulation (QAM), and Minimum Shift Keying (MSK) are the carrier modulation techniques that are dealt with in this chapter.

Chapter 4 elucidates the concept of analog-to-digital conversion. Sampling theorem and pulse modulation techniques are discussed first, followed by quantization and encoding techniques.

Chapter 5 focuses on data communication. The interface standards for serial and parallel communication are described. Furthermore, this chapter also explicates the error detection and correction techniques employed in data communication.

The basic limits of communication, including the information content of a memory-less source and the capacity of the Additive White Gaussian Channel (AWGN), form the basis of Chapter 6. The Shannon Fano coding algorithm, Huffman coding algorithm, and the linear block coding and convolutional codes are also illustrated in this chapter.

A handful of selected areas of communication like the fundamentals of mobile communication, cellular concepts, satellite communication, body area networks, and multiple access schemes are emphasized on in Chapter 7.

The worked out examples in this book are quite extensive and exemplify the techniques that are theoretically discussed. We hope that this book will be useful to communication engineers.

Acknowledgments

We are grateful to *Professor Zhi Ding, University of California Davis*, who consented to review our book and the River Publication team for their valuable insights and suggestions, which have greatly contributed to significant improvements of this text.

List of Figures

List of Tables

List of Abbreviations

AM	Amplitude Modulation
AMPS	Advanced Mobile Phone System
ANSI	American National Standards Institute
ASK	Amplitude Shift Keying
AWGN	Additive White Gaussian Noise
BFSK	Binary Frequency Shift Keying
BPSK	Binary Phase Shift Keying
BSC	Base Station Controller
BSS	Base Station Subsystem
BTS	Base Transceiver Station
CDMA	Code Division Multiple Access
CRC	Cyclic Redundancy Check
DCE	Data Communication Equipment
DSB-FC	Double Sideband Full Carrier Modulation
DSB-SC	Double Sideband Suppressed Carrier Modulation
DTE	Data Terminal Equipment
EIA	Electronic Industry Association
FDMA	Frequency Division Multiple Access
FM	Frequency Modulation
FSK	Frequency Shift Keying
GSM	Global System for Mobile
HLR	Home Location Register
IAB	Internet Activities Board
IEEE	Institute of Electrical and Electronic Engineers Information
ISO	International Organization for Standardization
ITU-T	International Telecommunication Union-Telecommunication
LRC	Longitudinal Redundancy Check
MSK	Minimum Shift Keying
NRZ	Non Return to Zero
PAM	Pulse Amplitude Modulation

PCM	Pulse Code Modulation
PM	Phase Modulation
PPM	Pulse Position Modulation
PSK	Phase Shift Keying
PSTN	Public Switched Telephone Network
PWM	Pulse Width Modulation
QAM	Quadrature Amplitude Modulation
QPSK	Quadrature Phase Shift Keying
RS-232-C	Recommended Standard-232-C
RZ	Return to Zero
SIM	Subscriber Identity Module
SNMP	Simple Network Management Protocol
SSB	Single Sideband Modulation
TDMA	Time Division Multiple Access
VLR	Visitor Location Register
VRC	Vertical Redundancy Check
VSB	Vestigial Sideband Modulation

PART I

Analog Communication

1

Analog Modulation

1.1 Introduction

Communication is the process of exchange of data or signal between two points. The two points are transmitter and a receiver. In both analog and digital communication, the transmitter conveys the information in the form of signals. Once we understand the concept of signals, we can easily classify into various categories of systems and understand their working.
Signal:

Signal is a physical quantity that varies with time or space or any other independent variable. Signals are classified into two types. They are:

1.1.1 Types of Signals

Continuous time signal:
The signal will be defined for all the time intervals. For example, $x(t) = \sin \omega t$. Here, the continuous signal $x(t)$ can be defined for all time intervals. The time period "t" varies from $-\infty < t < \infty$. Figure 1.1 illustrates the continuous time signal.

Discrete time signal:
A discrete time signal can be defined as only for a specific time interval. Figure 1.2 shows the discrete time signal. For example,

$$x(n) = \begin{cases} 1 & \text{when } n = 0 \\ 0 & \text{otherwise} \end{cases}.$$

Here, the discrete time "n" should be an integer, i.e. the amplitude of the signal can be defined as relevant to a time period $n = 1$ or 2 and so on. The signal cannot be defined for a time period $n = 0.5$, 1.5, and so on.

3

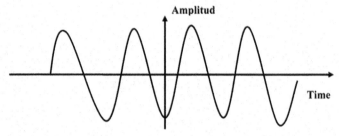

Figure 1.1   Continuous time signal.

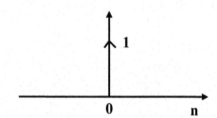

Figure 1.2   Discrete time signal.

1.2 Types of Communication

Based on the classification of signals, the communication systems can be classified as *Analog and digital communication systems.* The analog communication system uses continuous time signal as input, whereas digital communication uses discrete time signal as input.

1.2.1 Basic Blocks of Communication Systems

The term "*communication*" is transfer of information from one place to another or between individuals. There is a transmitter on one side and the receiver on the other. In between these, there is the channel as a medium to transmit information. It is shown in Figure 1.3.

Figure 1.3 can be elaborated through a look at communication systems, which are shown in Figure 1.4.

1.2.2 Detailed View of Communication Systems

A transmitter block is divided into source, source encoder, channel encoder, and modulator. Similarly a receiver block is divided into demodulator, channel decoder, source decoder, and detector.

Source: It is responsible for generating information, which may be continuous or discrete in nature.

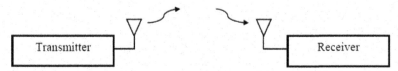

Figure 1.3 Basic blocks of communication systems.

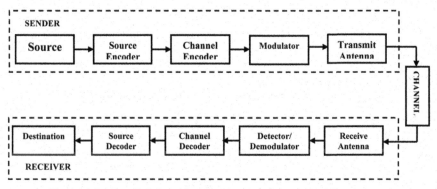

Figure 1.4 Detailed view of various blocks of communication systems.

Source encoder: It is an algorithm responsible for removing the redundant portion in information. For example, let us take the source input or data are represented by 0000 1111 0000 1111 1111 0000 0000. Then, the source encoder algorithm converts the input data into 0 1 0 1 1 0 0, i.e. a group of consecutive four zero bits or one bit is represented by 0 or 1. The purpose of the source encoder is bandwidth efficient data transmission, the bandwidth requirement to transmit my information is very large in the absence of a source encoder (i.e. 28 bits are transmitted but, when source encoding algorithm is used, only 7 bits are transmitted). So, automatically, the amount of bandwidth requirement is small. The detailed discussion on source encoder is given in Chapter 6.

Channel encoder: The purpose of the channel encoder is to add some extra bits to data and to have the data in more robust against noises. The detailed discussion on channel encoder is given in Chapter 7.

Modulator: The need for modulation has to be understood prior to that.

1.3 Need for Modulation

(i) To separate the various user signals or separate the signal from various transmitters

Audio signals are known to occupy the frequency range from 20 Hz to 20 kHz in the same frequency band. At the receiver side, picking up particular user signals is not possible, since all the user frequencies are occupied in the same frequency band. Hence, modulation is necessary for separation of various user signals. For example, the first user frequency is found to be occupied in between 20 Hz and 20 kHz, and it is modulated with a 3-MHz carrier signal. User-2 signal frequency is also found to occupy the same frequency band, but the carrier frequency is 5 MHz. It informs that, after the modulation process, user-1 signals travel in 3-MHz carrier signal and user-2 signals travels by 5-MHz carrier. At the receiver side, to pick user-1 signal, the receiver is tuned to 3 MHz and it receives user-1 signal alone. Hence, modulation is necessary for separation of various user transmitter signals.

(ii) Size of the antenna

Transmitting antenna size should be at least one quarter of wavelength of transmitting signal to ensure efficient transmission of the signal. For example, let us consider the modulation signal frequency as 15 kHz. To transmit the signal through antenna, the length of antenna should be 20 km in size. ($\lambda = c/f$; where c is the velocity of light in free space and its value is 3×10^8 m/ sec and f is frequency of operation. Hence, $\lambda = 3 \times 10^8/15 \times 10^3 = 20$ km). But when we do modulation, the size of the antenna is reduced. Let us take the signal frequency to be 15 kHz modulated by a 1-MHz carrier signal. Now, the user information travels in 1-MHz carrier signal. The size of the antenna required for this case is only 75 m ($\lambda = 3 \times 10^8/1 \times 10^6 = 75$ m). Hence, the modulation process is required for reduction of the size of the antenna.

(iii) Effective power transmission

Power radiated from the antenna is given as $P = (l/\lambda)^2$. It shows the ability of the antenna to transmit greater volume of power when it is small. The size of the antenna is known to be smaller only when the modulation process is considered.

Channel: The channel may be a wired medium or wireless medium one. Examples for wired lines are co-axial cable, optical fiber cable, and air as a medium for wireless transmission.

1.4 Modulation

The basic definition is: Changing the characteristic of the carrier wave in accordance with the amplitude of the modulating signal or the message signal or the information signal.

Carrier signal characters are *amplitude, frequency, and phase angle*. Hence, the amplitude of the carrier or frequency of the carrier or phase of the carrier can be varied with respect to the amplitude of the message signal.

Hence, the corresponding modulation techniques are: Amplitude modulation (AM), Frequency modulation (FM), and Phase modulation (PM).

A message signal is a low-frequency signal, while a carrier signal is a high-frequency signal. Message and carrier signals are needed for doing modulation. The output of the modulator is modulated signal.

1.4.1 Amplitude Modulation

Definition:

This involves changing the amplitude of the carrier in accordance with the amplitude of the message signal. Figure 1.5 illustrates AM waveform.

Mathematical analysis:

Let us take modulating signal as

$$e_m = E_m \cos 2\pi f_m t \tag{1.1}$$

where e_m represents the message signal, E_m represents the amplitude of the message signal, and f_m is the frequency of the modulating signal.

Similarly, the carrier signal is denoted by

$$e_c = E_c \cos 2\pi f_c t \tag{1.2}$$

where e_c represents the carrier signal, E_c represents the amplitude of carrier signal, and f_c is the carrier frequency.

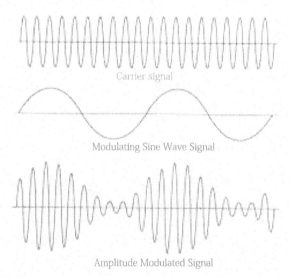

Carrier signal

Modulating Sine Wave Signal

Amplitude Modulated Signal

Figure 1.5 Time domain representation of AM signal.

Then the modulated signal is given by

$$e_{AM} = [E_c + E_m \cos 2\pi f_m t] \cos 2\pi f_c t \tag{1.3}$$

$$= E_c \left[1 + \frac{E_m}{E_c} \cos 2\pi f_m t\right] \cos 2\pi f_c t \tag{1.4}$$

1.4.1.1 Modulation index (m)

It is the ratio of the amplitude of the message signal to the amplitude of the carrier signal.

$$m = \frac{E_m}{E_c} \tag{1.5}$$

Hence, Equation (1.4) becomes

$$e_{AM} = E_C [1 + m \cos 2\pi f_m t] \cos 2\pi f_c t$$

Now multiply $E_c \cos 2\pi f_c t$ with the contents inside the bracket.

$$e_{AM} = E_c \cos 2\pi f_c t + m E_c \cos 2\pi f_c t \cos 2\pi f_m t \tag{1.6}$$

Apply $\cos A \cos B = \frac{1}{2} [\cos (A + B) + \cos (A - B)]$ formula to the second term in the summation

$$e_{AM} = E_c \cos 2\pi f_c t + m E_c / 2 \left\{\cos 2\pi (f_c + f_m) t + \cos 2\pi (f_c - f_m) t\right\} \tag{1.7}$$

This is a signal made up of three signal components

- A carrier at frequency f_c Hz
- Upper side frequency at $f_c + f_m$ Hz and
- Lower side frequency at $f_c - f_m$ Hz

The **bandwidth** (the difference between the highest and the lowest frequencies) is

$$\text{Bandwidth} = (f_c + f_m) - (f_c - f_m) = 2 f_m \text{ Hz} \tag{1.8}$$

where f_m is the maximum frequency of the modulating signal.

The time domain and frequency spectrum of AM signal can be drawn from Equation (1.7). It is shown in Figures 1.5 and 1.6.

Figure 1.6 suggests that, frequency spectrum consists of

(i) $E_c \cos 2\pi f_c t$-carrier frequency f_c with the amplitude of E_c

(ii) side band at $(f_c + f_m)$ and $(f_c - f_m)$ with the amplitude of $m E_c / 2$.

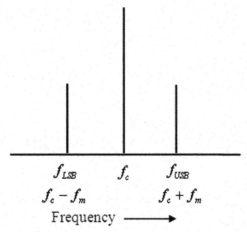

Figure 1.6 Frequency domain display of AM signal.

The relationship between time and frequency domain is shown in Figure 1.7.

From Figure 1.7, it is clear that modulated signal in time domain is only the aggregate of the modulating signal, the carrier signal, the upper sideband signals, and the lower sideband signal.

In other words, AM wave is the algebraic sum of the carrier, the upper and the lower sideband frequencies. It is shown in Figure 1.8.

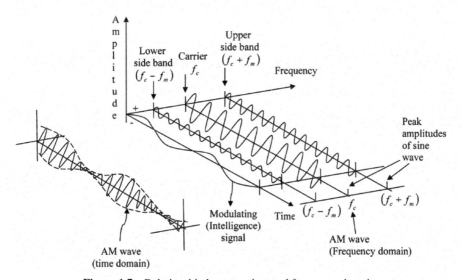

Figure 1.7 Relationship between time and frequency domain.

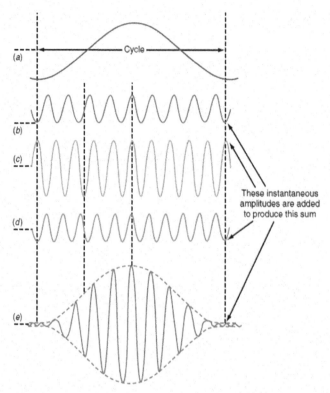

Figure 1.8 Where (*a*) modulating signal (*b*) carrier signal (*c*) and (*d*) upper and lower sideband signals and (*e*) composite AM wave.

1.5 Modulation of Complex Modulating Signal

In the previous analysis, simple cosine or sinusoidal signal has been considered as the message signal. In real-life scenario, the modulating signal is like a complex signal. In other words, the modulating signal is the sum of two or more cosine or sinusoidal signals. Hence, the method of modulation of a complex signal with a high-frequency carrier is discussed. What is the spectrum for modulated signal? What is the bandwidth? What is the modulation index for complex modulating signal? These are questions to be answered.

Let us consider modulating signal $x(t)$ as, sum of two signals, i.e.

$x(t) = x_1(t) + x_2(t)$ where $x_1(t)$ and $x_2(t)$ are given as,

$x_1(t) = E_{m1} \cos 2\pi f_{m1} t$ and $x_2(t) = E_{m2} \cos 2\pi f_{m2} t$, where E_{m1} is the amplitude of $x_1(t)$ and f_{m1} is the frequency of $x_1(t)$. Similarly, E_{m2} is the amplitude of $x_2(t)$ and f_{m2} is the frequency of $x_2(t)$.

Now the carrier signal is represented as

$e_c = E_c \cos 2\pi f_c t$, where e_c represents the carrier signal, E_c represents the amplitude of carrier signal, and f_c is the carrier frequency.

Then the modulated signal is given as,

$$e_{AM} = E_{AM} \cos 2\pi f_c t,$$

Here, the amplitude of the modulated signal E_{AM} is given as,

$E_{AM} = E_c + E_{m1} \cos 2\pi f_{m1} t + E_{m2} \cos 2\pi f_{m2} t$. Now, substituting the value of E_{AM} in the above equation, we get

$$e_{AM} = (E_c + E_{m1} \cos 2\pi f_{m1} t + E_{m2} \cos 2\pi f_{m2} t) \cos 2\pi f_c t$$
$$= E_c \left(1 + \frac{E_{m1}}{E_c} \cos 2\pi f_{m1} t + \frac{E_{m2}}{E_c} \cos 2\pi f_{m2} t\right) \cos 2\pi f_c t$$

Now, Modulation index 1 = $m_1 = \frac{E_{m1}}{E_c}$

Modulation index 2 = $m_2 = \frac{E_{m2}}{E_c}$

Hence, the total modulation index is $m = \sqrt{m_1^2 + m_2^2}$

Therefore, the modulated signal can written as

$$e_{AM} = E_c \left(1 + m_1 \cos 2\pi f_{m1} t + m_2 \cos 2\pi f_{m2} t\right) \cos 2\pi f_c t$$

$$e_{AM} = (E_c \cos 2\pi f_c t + m_1 E_c \cos 2\pi f_c t \cos 2\pi f_{m1} t$$
$$+ m_2 E_c \cos 2\pi f_c t \cos 2\pi f_{m2} t)$$

Apply the Cos A and Cos B formula to $\cos 2\pi f_c t \cos 2\pi f_{m1} t$ and $\cos 2\pi f_c t \cos 2\pi f_{m2} t$

$$\cos A \cos B = \frac{\cos (A + B) + \cos (A - B)}{2}$$

So, $\cos 2\pi f_c t \cos 2\pi f_{m1} t$ becomes,

$$\frac{\cos 2\pi (f_c + f_{m1}) t + \cos 2\pi (f_c - f_{m1}) t}{2}$$

Therefore,

$$e_{AM} = \left(E_c \cos 2\pi f_c t + \frac{m_1 E_c}{2} [\cos 2\pi (f_c + f_{m1}) t + \cos 2\pi (f_c - f_{m1}) t]\right.$$
$$\left. + \frac{m_2 E_c}{2} [\cos 2\pi (f_c + f_{m2}) t + \cos 2\pi (f_c - f_{m2}) t]\right)$$

The final modulated signal in frequency domain is implied as consisting of

1. Carrier signal $E_c \cos 2\pi f_c t$
2. Upper sideband at $f_c + f_{m1}$ and $f_c + f_{m2}$
3. Lower sideband at $f_c - f_{m1}$ and $f_c - f_{m2}$

Frequency spectrum is shown in Figure 1.9.

Total bandwidth for AM when the input signal is a complex modulating signal Bandwidth = $2 \times f_{m(\max)}$

where $f_{m(\max)}$ is the maximum frequency of modulating signal.

1.6 Importance of Modulation Index

Modulation index is known as just the ratio of modulating signal amplitude to the carrier signal amplitude,

$$m = \frac{E_m}{E_c}$$

Modulation index should be in the range of $0 < m < 1$.

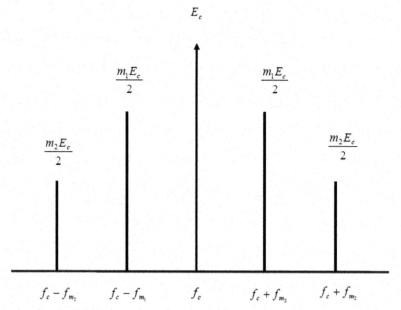

Figure 1.9   Frequency spectrum of complex modulating signal.

Based on the values of m, the AM modulation can be classified into two types. They are:

1. Linear modulation and
2. Over modulation.

Linear modulation:

The type of modulation is linear modulation when $m \leq 1$. For proper signal reception at receiver side, $m \leq 1$ should be maintained. The corresponding waveform is shown in Figure 1.10.

Over modulation:

The type of modulation is over modulation when $m > 1$. Upper and lower envelopes are combined with each other due to $m > 1$ and so there may be a

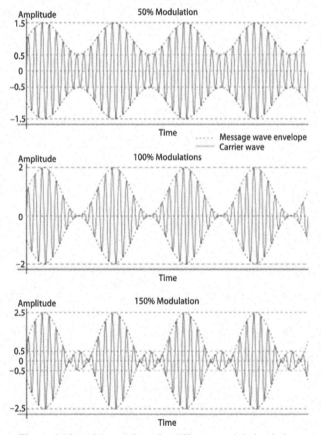

Figure 1.10 AM waveform for different modulation index.

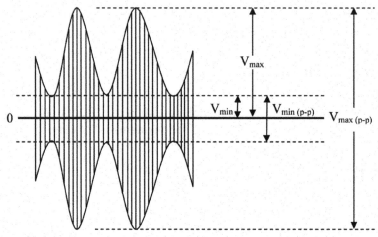

Figure 1.11 AM waveform for calculation of modulation index m.

phase reversal in the modulated signal. Perfect reconstruction is therefore not possible in over modulation.

In Figure 1.10, above a and b have modulation indices of 0.5 and 1 respectively, figure c has modulation index of 1.5. Hence, phase reversal can be seen from the figure.

Another way to represent modulation index m:

$$m = \frac{V_{\max} - V_{\min}}{V_{\max} + V_{\min}},$$

where E_m or $V_m = V_{\max} - V_{\min}$ and E_c or $V_c = V_{\max} + V_{\min}$, $V_{\max}$ and $V_{\min}$ are shown in Figure 1.11.

1.6.1 Depth of Modulation or Percent Modulation

It is modulation index m multiplied by 100 ($m \times 100$).

1.6.2 Transmission Efficiency of AM Modulation

It is the ratio of power in sidebands to total power.

$$\eta = [\text{Power in sidebands/Total power}] \times 100$$

$$\eta = \frac{P_{\text{USB}} + P_{\text{LSB}}}{P_T} \times 100$$

Transmission efficiency (η) can relate with modulation index (m) by

$$\eta = \frac{m^2}{2 + m^2} \times 100$$

1.6.3 AM Power Calculation

Total transmitted power (P_T) is the sum of carrier power (P_C), the power in the upper side band (P_{USB}) and the power in the lower side band (P_{LSB}).

$$P_T = P_C + P_{USB} + P_{LSB}$$

$P_T = \frac{E^2_{CARR}}{R} + \frac{E^2_{USB}}{R} + \frac{E^2_{LSB}}{R}$; Where $E_{CARR}, E_{USB}, E_{LSB}$ are RMS values of the carrier and sideband amplitude and R is the characteristic impedance of antenna in which the total power is dissipated.
The carrier power is given by,
$P_C = \frac{E^2_{CARR}}{R} = \frac{\left[E_C/\sqrt{2}\right]^2}{R} = \frac{E^2_C}{2R}$; Where E_C is peak carrier amplitude.
In similar manner, power in side band is given by,

$$P_{USB} = P_{LSB} = \frac{E^2_{SB}}{R}$$

As we know the peak amplitude of each sideband is $\frac{mE_C}{2}$

$$P_{USB} = P_{LSB} = \frac{E^2_{SB}}{R} = \frac{\left[mE_C/2\sqrt{2}\right]^2}{R} = \frac{m^2 E^2_C}{8R}$$

$$P_{USB} = P_{LSB} = \frac{m^2}{4} \times \frac{E^2_C}{2R}$$

As we know $P_C = \frac{E^2_C}{2R}$
Therefore,

$$P_{USB} = P_{LSB} = \frac{m^2}{4} \times P_C$$

Total power is, $P_T = P_C + P_{USB} + P_{LSB}$

$$P_T = P_C + \frac{m^2}{4} \times P_C + \frac{m^2}{4} \times P_C$$

$$P_T = P_C + \frac{m^2}{2} \times P_C$$

The total power in terms of modulation index can be expressed by,

$$P_T = P_C \left(1 + {m^2}/{2}\right)$$

The current equation for AM can be found in the same way.

$$P_T = P_C \left(1 + \frac{m^2}{2}\right)$$

$$I_T^2 = I_C^2 \left(1 + {m^2}/{2}\right)$$

$$I_T = I_C \sqrt{\left(1 + \frac{m^2}{2}\right)}$$

When the carrier power of AM broadcast station is 1000 W and the modulation index is 1 or percent modulation is 100, the total power is

$$P_T = 1000 \left(1 + \frac{1}{2}\right)$$

$$P_T = 1500 \text{ W}.$$

It implies, carrier power is 1000 W and side band power is 500 W (1000 W + 500 W = 1500 W). Hence, power in the individual sideband is 250 W.

Hence, total sideband power is half of carrier power for 100% modulation.

When the modulation index is less than one, power consumed by the side bands is not half of the carrier power, but only a small volume of power consumed in sidebands.

For example, $m = 0.7$ and carrier power is 1000 W, then total power is given by

$$P_T = 1000 \left(1 + \frac{(0.7)^2}{2}\right) = 1245 \text{ W}.$$

It informs carrier power as 1000 W and the remaining 245 W is the sideband power. The sideband powers, it may be noted, are not half of the carrier power. This shows the importance of modulation index.

The spectrum of AM signal leads to the conclusion that, in the conventional AM or Double sideband full carrier (DSB-FC) technique, the total power transmitted from the radio station is mostly wasted in carrier, since the carrier does not convey any useful information. Also, both the sidebands carry the same information. So power in one sideband is also wasted. Power inefficiency is the major problem for DSB-FC requiring introduction of various AM techniques.

1.6.4 DSB-SC-Double Sideband Suppressed Carrier

Carrier is known to occupy most of the transmission power and also not conveying any useful information. Hence, the idea is to suppress the carrier and transmit only the side bands, since sidebands alone convey the information signal f_m.

Figures 1.12 and 1.13 illustrate the time domain and frequency domain view of DSB-SC signals, respectively.

Total power in DSB-SC:

$$P_{T(\text{DSB}-\text{SC})} = P_C {}^{\text{m}^2}\!/_2$$

Time domain representation of DSB-SC signals:

The envelope of the signal is not the same as that of the modulation signal. The unique characteristic of DSB-SC signal is phase transition that occurs at the lower amplitude portion of the wave.

Suppression of the carrier may be seen in the frequency domain. Only the sideband alone is transmitted. DSB-SC signals are generated by a balanced modulator.

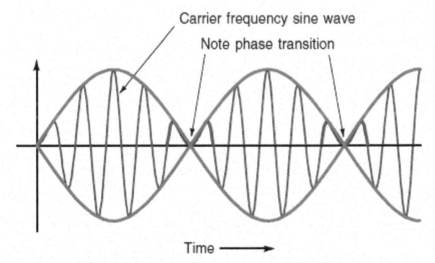

Figure 1.12   Time domain representation of DSB-SC signal.

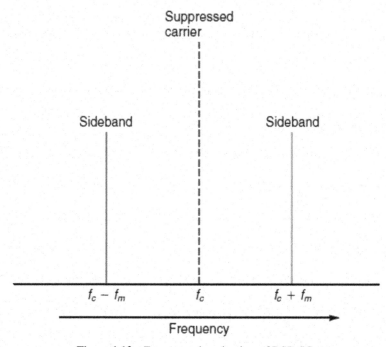

Figure 1.13   Frequency domain view of DSB-SC.

The DSB-SC does not find extensive use despite the suppression of the carrier in the technique. This is because demodulation requires a very complex circuit. One important application of DSB-SC technique is in the conveying of color information TV broadcasting.

In DSB-SC, both the side bands covey the same information. Hence, the next possible idea is to eliminate one of the sidebands, with the resulting technique being a single sideband modulation (SSB).

1.6.5 SSB-Single Sideband Modulation

The carrier and one of the sidebands are suppressed in SSB; only one sideband is transmitted from radio station. Time domain and frequency domain view of SSB signals are noted in Figures 1.14 and 1.15.

Total power in SSB technique:

$$P_{T(\text{SSB})} = P_C \frac{m^2}{4}.$$

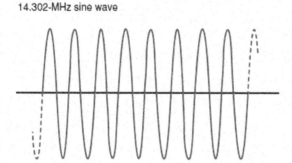

SSB signal
14.302-MHz sine wave

Figure 1.14 Time domain representation of SSB.

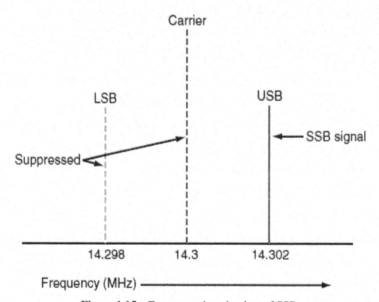

Figure 1.15 Frequency domain view of SSB.

SSB technique is more power efficient compared with other schemes, considering the transmission of the carrier signal in the conventional DSB-FC, the radio station despite the absence of any modulating or message signal. But in SSB technique, sideband is produced only in the presence of the modulating signal. This is the reason why SSB is more power efficient than other schemes.

In this figure, modulating signal is 2 kHz and carrier frequency is 14 MHz.

The main limitation for SSB technique is the difficulty in detection of the signal at the receiver side.

The main application for the SSB technique is the two-way radios. Two-way SSB communications are used in marine applications.

1.6.6 VSB-Vestigial Sideband Modulation

VSB modulation technique is normally used in TV Broadcasting. The TV signal consists of a picture and an audio signal. Both occupy at different frequency regions, i.e. audio carriers are frequency modulated and video information is amplitude modulated.

Video information is typically available in the region of 4.2 MHz. Hence, the total bandwidth is 8.4 MHz. But, according to the FCC standard, TV bandwidth is 6 MHz. Hence, to make 8.4 MHz bandwidth to 6 MHz, some of the video carriers are suppressed (or small portions or vestige), and the resulting modulation is VSB. Normally video signals above 0.75 MHz (750 kHz) are suppressed in the lower sideband and upper sidebands are completely transmitted. Figure 1.16 shows the VSB technique.

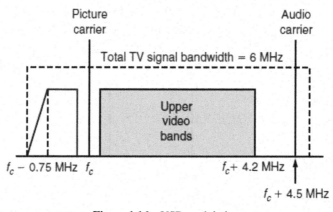

Figure 1.16   VSB modulation.

1.7 Comparison of Various AM Modulation Technique

Following Table 1.1 lists the comparison of various AM modulation schemes.

Table 1.1   Comparison of various AM modulation techniques

S.No.	Parameter	DSB-FC	DSB-SC	SSB
1	Carrier suppression	Not available	Fully	Fully
2	Sideband suppression	Not available	Not available	One sideband completely suppressed

(Continued)

Table 1.1 Continued

3	Bandwidth	$2f_m$	$2f_m$	f_m
4	Transmission Efficiency	Minimum	Moderate	Maximum
5	Number of modulating input	1	1	1
6	Application	Radio broadcasting	Radio broadcasting	Point-to-point mobile communication
7	Power requirement	High	Medium	Very small
8	Power saving for sinusoidal	33.33%	66.66%	83.3%
9	Power saving for non-sinusoidal	33.33%	50%	75%
10	Complexity	Simple	Simple	Complex

1.8 Solved Problems

1. *A 400 W carrier is modulated to a depth of 75%. Calculate the total power in the modulated wave?*

   Given:
   Carrier Power $P_c = 400\,\text{W}$,
   Depth of modulation = 75%, hence modulation index $m = 0.75$
   Answer:
   Total power in the modulated wave,

   $$P_T = P_C \left(1 + \frac{m^2}{2}\right)$$
   $$= 400 \left(1 + \frac{0.75^2}{2}\right)$$
   $$P_T = 512.4\,\text{W}$$

2. *A broadcast radio transmitter radiates 10 kW when the modulation percentage is 60. Calculate the carrier power.*

   Given:
   $$P_T = 10\,\text{kW} = 10 \times 10^3$$

   Modulation index $(m) = 0.6$
   Answer:
   We know, $P_T = P_C \left(1 + \frac{m^2}{2}\right)$
   Then,
   $$P_C = \frac{P_T}{(1 + m^2/2)}$$
   $$= \frac{10 \times 10^3}{\left(1 + \frac{0.6^2}{2}\right)}$$

$$P_C = 8.47kW$$

3. *A* 1 MHz *carrier with an amplitude of* 1 V *peak is modulated by a* 1 kHz
   signal with modulation index 0.5. *Sketch the frequency spectrum.*

Given:

Carrier frequency $f_c = 1\,\text{MHz}$

Carrier amplitude $E_c = 1\,\text{V}$

Message signal frequency $f_m = 1\,\text{kHz}$

Answer:

Upper side band component:

$$f_c + f_m = 1000\,\text{kHz} + 1\,\text{kHz} = 1001\,\text{kHz}$$

Upper sideband amplitude $= \frac{mE_C}{2}$
$$= \frac{0.5 \times 1}{2} = 0.25 \text{ V}$$

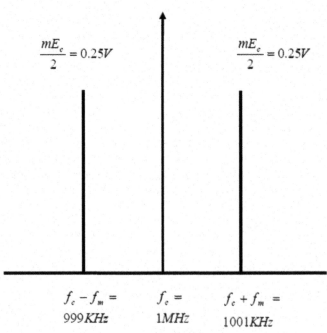

Figure 1.17   Frequency spectrum.

Lower sideband component:

$$f_c - f_m = 1000\,\text{kHz} - 1\,\text{kHz} = 999\,\text{kHz}$$

Lower sideband amplitude = $\frac{mE_C}{2}$

$$= \frac{0.5 \times 1}{2} = 0.25\,\text{V}$$

Hence, the frequency spectrum is shown in Figure 1.17.

4. *AM radio channel bandwidth is* 10 kHz. *What is the maximum modulation frequency?*

   Given: channel bandwidth (B) = 10 kHz.
   Answer:
   Bandwidth of AM signal $= 2f_m$

$$f_m = \frac{\text{Bandwidth}}{2}$$

$$= 10\,\text{kHz}/2 = 5\,\text{kHz}$$

5. *For an AM DSB-FC transmitter with an un-modulated carrier power* $P_c = $ 100 W *that is modulated simultaneously by three modulating signals with coefficient of modulation* $m_1 = 0.2, m_2 = 0.4$ *and* $m_3 = 0.5$, *determine,*

   (a) *Total coefficient of modulation*
   (b) *Upper and lower sideband power*
   (c) *Total transmitted power*

   Given:
   Carrier power: $P_c = 100\,\text{W}$
   Modulation indices: $m_1 = 0.2, m_2 = 0.4$ and $m_3 = 0.5$
   Answer:

   a) Modulation index $(m) = \sqrt{\left(m_1^2 + m_2^2 + m_3^2\right)}$

$$= \sqrt{\left(0.2^2 + 0.4^2 + 0.5^2\right)}$$

$$m = 0.67$$

   b) Upper and lower sideband power

$$P_{\text{USB}} = P_{\text{LSB}} = P_C \frac{m^2}{4}$$

$$= 100 \times \frac{0.67^2}{4}$$
$$= 11.25\,\text{W}$$

c) Total transmitted power

$$P_T = P_C \left(1 + m^2/2\right)$$

$$= 100 \left(1 + \frac{0.67^2}{2}\right)$$

$$P_T = 122.45W$$

6. *What is the efficiency of AM system, when the modulation index is one?*

$$\eta = \frac{m^2}{2 + m^2} \times 100$$

$$\eta = \frac{1}{2 + 1} \times 100 = \frac{1}{3} \times 100 = 33.33\%$$

7. *A Modulating signal* $20 \sin \left(2\pi \times 10^3 t\right)$ *is used to modulate a carrier signal* $40 \sin \left(2\pi \times 10^4 t\right)$. *Find out,*

(a) *Modulation index*
(b) *Percentage modulation*
(c) *Frequencies of sideband and their amplitudes*
(d) *Bandwidth of modulating signal*
(e) *Draw the spectrum of AM wave.*

Given:
Modulating signal(e_m): $20 \sin \left(2\pi \times 10^3 t\right)$
Carrier signal (e_c); $40 \sin \left(2\pi \times 10^4 t\right)$
Given informations are compared with generalized expression of modulating and carrier signal.

$$e_m = E_m \cos 2\pi f_m t$$

$$E_m = 20; f_m = 10^3$$

$$e_c = E_c \cos 2\pi f_c t$$

$$E_c = 40; f_c = 10^4$$

a) Modulation index (m)

$$m = \frac{E_m}{E_c}$$

$$m = \frac{20}{40} = 1/2 = 0.5$$

b) Percentage of modulation $= m \times 100$

$$= 0.5 \times 100 = 50\%$$

c) Frequencies of sidebands
   Upper sideband $f_c + f_m = 10^4 + 10^3 = 10\,\text{kHz} + 1\,\text{kHz} = 11\,\text{kHz}$
   Lower sideband $f_c - f_m = 10\,\text{kHz} + 1\,\text{kHz} = 9\,\text{kHz}$
   Sideband amplitudes $= \frac{mE_C}{2}$

$$= \frac{0.5 \times 40}{2} = 10V$$

d) Bandwidth $= 2f_m$

$$= 2 \times 1\,\text{kHz} = 2\,\text{kHz}$$

e) Frequency spectrum

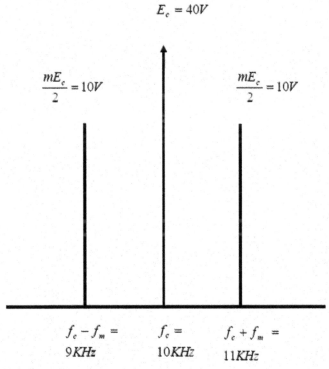

Figure 1.18 Frequency spectrum.

8. *A complex modulating signal consisting of a sine-wave of amplitude* 3 V *peak and frequency* 1 kHz, *and a cosine wave of amplitude* 5 V *and frequency* 3 kHz *modulates a* 500 kHz *and* 10 V *peak carrier voltage. Plot the spectrum of AM signal.*

Given:

Modulating signal:

$$E_{m1} = 3\,\text{V}; f_{m1} = 1\,\text{kHz}$$

$$E_{m2} = 5\,\text{V}; f_{m2} = 3\,\text{kHz}$$

Carrier signal:

$$E_c = 10\,\text{V}; f_c = 500\,\text{kHz}$$

Answer:

$$m_1 = \frac{E_{m1}}{E_c} = \frac{3}{10} = 0.3$$

$$m_2 = \frac{E_{m2}}{E_c} = \frac{5}{10} = 0.5$$

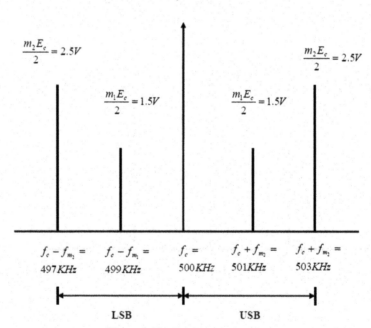

Figure 1.19   Frequency spectrum.

Total modulation index: $m = \sqrt{(m_1^2 + m_2^2)}$

$$m = \sqrt{(0.3^2 + 0.5^2)} = 0.583$$

Upper sideband components;

$$f_c + f_{m1} = 500\,\text{kHz} + 1\,\text{kHz} = 501\,\text{kHz}$$

$$f_c + f_{m2} = 500\,\text{kHz} + 3\,\text{kHz} = 503\,\text{kHz}$$

Lower sideband components;

$$f_c - f_{m1} = 500\,\text{kHz} - 1\,\text{kHz} = 499\,\text{kHz}$$

$$f_c - f_{m2} = 500\,\text{kHz} - 3\,\text{kHz} = 497\,\text{kHz}$$

Bandwidth

$$B = 2 \times f_{m(\text{max})}$$
$$= 2 \times 3\,\text{kHz} = 6\,\text{kHz}$$

2

Angle Modulation

2.1 Mathematical Analysis

The process of changing the angle of the carrier signal in accordance with the amplitude of the message signal is referred as angle modulation.

The angle modulated signal is represented as

$$x(t) = A \cos [\omega_c t + \varphi(t)] \tag{2.1}$$

where A is the amplitude of angle modulated signal, $\omega_c = 2\pi f_c$, f_c is the carrier frequency and is constant, and $\varphi(t)$ is the phase information and it is a time variant quantity. In angle modulation, $\varphi(t)$ carries the information to be conveyed from the transmitter to the receiver, i.e. modulating signal modifies the $\varphi(t)$ in appropriate manner.

Now let us take the argument of cosine function as

$$\theta_i(t) = \omega_c t + \varphi(t) \tag{2.2}$$

where $\theta_i(t)$ is the instantaneous phase of the carrier. The question arises as what is instantaneous frequency. It is only a derivative of instantaneous phase.

Hence, instantaneous frequency $\omega_i(t)$ is,

$$\omega_i(t) = \frac{d\vartheta_i(t)}{dt} = \omega_c + \frac{d\phi(t)}{dt} \tag{2.3}$$

Here, $\varphi(t)$ is the instantaneous phase deviation. Then, instantaneous frequency deviation is a derivative of instantaneous phase deviation.

Therefore, instantaneous frequency deviation $d\phi(t)/dt$.

These definitions lead us to those of what is PM and FM?

PM is,

$$\varphi(t) = k_p m(t) \tag{2.4}$$

where k_p is the phase modulation constant.

Similarly, FM is

$$\frac{d\phi(t)}{dt} = k_f m(t) \tag{2.5}$$

29

where k_f is the frequency modulation constant. So, phase $\varphi(t)$ for FM is found through integration on both sides of the above equation.

$$\text{Hence, } \phi(t) = k_f \int_{-\infty}^{t} m(s)\,ds \tag{2.6}$$

Therefore, PM signal is given as

$$x_{\text{PM}}(t) = A\cos[\omega_c t + k_p m(t)] \tag{2.7}$$

For FM,

$$x_{\text{FM}}(t) = A\cos\left[\omega_c t + k_f \int_{-\infty}^{t} m(s)\,ds\right] \tag{2.8}$$

Where the message signal or modulating signal $m(t)$ is, $m(t) = A_m \cos \omega_m t$.
So, integration of $m(t)$ is $\frac{A_m}{\omega_m}\sin \omega_m t$. Substituting this value into generalized expression of FM,

$$x(t) = A\cos\left[\omega_c t + k_f \frac{A_m}{\omega_m}\sin \omega_m t\right] \tag{2.9}$$

Now, the modulation index of FM is β, and it is given as

$$\beta = k_f \frac{A_m}{\omega_m} \tag{2.10}$$

Another way to express β is,

$$\beta = \frac{\Delta f}{f_m} \tag{2.11}$$

where Δf is the maximum frequency deviation and f_m is the maximum frequency of modulating signal frequency.
Hence, the finalized expression of FM is

$$x(t) = A\cos[\omega_c t + \beta \sin \omega_m t] \tag{2.12}$$

The above expression is in the form of cos [... .sin] format, i.e. periodic function (....periodic function). These types of expressions are solved with the help of Bessel functions. Hence, the final expression is

$$x_{FM}(t) = A\begin{Bmatrix} [J_0(\beta)\sin \omega_c t] + J_1(\beta)\,[\sin(\omega_c + \omega_m)t \\ -\sin(\omega_c - \omega_m)\,t] + J_2(\beta)\,[\sin(\omega_c + 2\omega_m)\,t \\ -\sin(\omega_c - 2\omega_m)t] + J_3(\beta)[\sin(\omega_c + 3\omega_m)\,t \\ -\sin(\omega_c - 3\omega_m)\,t] + \ldots\ldots\ldots \end{Bmatrix} \tag{2.13}$$

i.e. FM signal frequency spectrum consists of carrier component + infinite number of side bands. The same can be identified from Figure 2.1, the frequency spectrum of FM signal.

2.1.1 Bandwidth Calculation of FM

Theoretically, the bandwidth occupied by FM is infinite, since it has infinite number of side bands. But there is a thumb rule available for finding bandwidth of FM, it is

$$\text{Bandwidth (BW)} = 2 \times f_m \times \text{number of significant sidebands} \quad (2.14)$$

i.e. all the infinite number of sidebands carry the useful information. Only the finite number of sidebands carry the useful information.

Another way to find bandwidth is the "Carson's rule" according to which

$$BW = 2\left[\Delta f + f_m\right] \quad (2.15)$$

This is the usual method for finding bandwidth of FM signal.

Now, we need to recall the modulation index (β) of FM, $\beta = \frac{\Delta f}{f_m}$, since both β and BW depend on Δf and f_m.

2.1.2 Types of FM

When β is less than 1, it is a narrow band FM and when $\beta \gg 1$, it is wide-band FM.

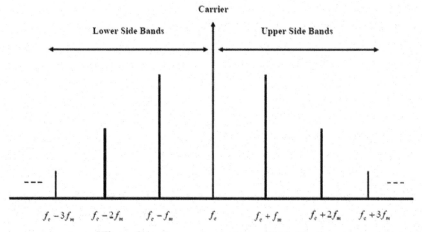

Figure 2.1 Frequency spectrum of FM signal.

Following Table 2.1 lists the comparison of narrow band and wide band FM schemes.

Table 2.1   Comparison of narrow band and wide band FM

S.No.	Parameter	Narrowband FM	Wideband FM
1	Modulation index (β)	Less than one	Greater than one
2	Maximum frequency deviation (Δf)	5 kHz	75 kHz
3	Range of modulating frequency signal	30 Hz to 3 kHz	30 Hz to 15 kHz
4	Maximum modulation index	1	5 to 2500
5	Bandwidth	Small, approximately same as AM	Large, it is around 15 times higher than narrowband FM

Total power requirement of FM:

$$P_T = \frac{E_c^2}{2R_L} \tag{2.16}$$

where R_L − load resistance.

Deviation ratio (D) or percentage of modulation of FM:

This is only the ratio of maximum frequency deviation to maximum modulating signal frequency,

$$D = \frac{\Delta f_{\mathrm{mac}}}{f_{m(\mathrm{max})}} \tag{2.17}$$

2.2 Mathematical Analysis of PM

$$x_{\mathrm{PM}}(t) = A \cos\left[\omega_c t + k_p m(t)\right] \tag{2.18}$$

Where $m(t) = A_m \cos \omega_m t$.
Hence,

$$x_{\mathrm{PM}}(t) = A \cos\left[\omega_c t + k_p A_m \cos \omega_m t\right] \tag{2.19}$$

Now modulation index of PM is,

$$\beta_{\mathrm{PM}} = k_p A_m \tag{2.20}$$

Once again for PM signal also, cos (....cos) function, i.e. periodic function (periodic function). Hence, the Bessel function expression is applied to the above equation and the final spectrum also consists of an infinite number of sidebands. The same can be noted in Figure 2.2.

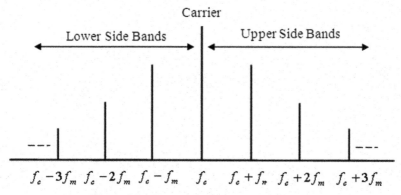

Figure 2.2 Frequency spectrum of PM signal.

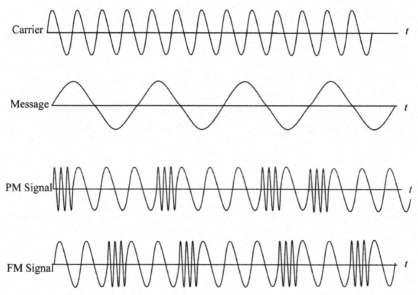

Figure 2.3 Time domain representation of FM and PM signal.

2.3 Noises

Noise is signal not required. The basic classification of noise is: "internal and external"

External Noise: This type of noise is man-made and arises from atmospheric environment.

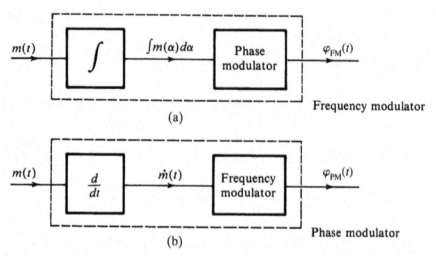

Figure 2.4   (a) Generation of FM using PM (b) Generation of PM using FM modulation.

Internal Noise: This type of noise arises from spontaneous fluctuation of voltage and current in an electric circuit. The most common examples of spontaneous fluctuation are shot noise and thermal noise.

Among noises of different types, internal noise causes a major problem to communication systems.

2.3.1 Types of Internal Noises

Shot noise or Poisson's noise:

It occurs in electronic devices such as diodes and transistors, due to the discrete nature of current flow in these devices.

Hence, electrons are emitted at random times denoted by τ_k, where k varies from $-\infty < k < \infty$.

Total current flow through the diode is the sum of all the current pulses. It is given by

$$X(t) = \sum_{k=-\infty}^{\infty} h(t - \tau_k) \tag{2.21}$$

where $h(t - \tau_k)$ is the current pulse generated at time τ_k.

The process $X(t)$ is called a stationary shot noise. The mean value for $X(t)$ is $\mu_X = \lambda \int_{-\infty}^{\infty} h(t)\, dt$, where λ is the rate of process and $h(t)$ is the waveform of current pulse.

Shot noise can be modeled using the Poisson process. Hence, shot noise is also referred to as Poisson's process.

Thermal noise or Johnson's noise or Nyquist noise:
Metallic conductor or resistor is known to contain a number of free electrons. Owing to thermal agitation, these free electrons move continuously in the conductor, causing collision to some other atoms and this process continues in nature at conductor.

These random motions of electrons in the conductor produces voltage fluctuations across the conductor. It causes noise voltage across the terminal.

Hence, thermal noise is

$$V_N^2 = 4\,\text{KBTR} \tag{2.22}$$

where K-Boltzmann's constant-$1.38 \times 10^{-23}\,J/K$, B–Equivalent noise bandwidth, T–Equivalent noise temperature, and R–Resistance across the circuit.

For example, $R = 1\,\text{k}\Omega$, $B = 5\,\text{MHz}$, $T = 290°$, then $V_N^2 = 9\mu v$.

Frequency spectrum of thermal noise is *"Flat"*, i.e. noise per unit bandwidth is constant.

Noise per unit bandwidth is termed as *Noise power spectrum density*.

Noise Power:
Noise power is given as

$$N_O = \frac{P_N}{B_N} \tag{2.23}$$

Noise temperature:
The concept associated with noise power is the effective noise temperature. It is given as

$$T_E = \frac{P_N}{K B_N} \tag{2.24}$$

Hence, $N_O = K T_E$ (relation between noise power and noise temperature).

Additive White Gaussian Noise (AWGN):
Noise of this type is added to noise of any type. White refers to uniform power across all frequency areas. Gaussian refers to white noise following normal distribution.

2.4 Solved Problems

1. *In a FM system, if the maximum value of deviation is* 75 kHz *and maximum modulating frequency is* 10 kHz. *Calculate the deviation ratio and find the bandwidth of the system using Carson's rule.*

   Given:
   Maximum value of deviation = 75 kHz
   Maximum modulating frequency = 10 kHz
   Answer:

   $$D = \frac{\Delta f_{\text{mac}}}{f_{m(\text{max})}}$$

   $$D = \frac{75 \text{ kHz}}{10 \text{ kHz}} = 7.5$$

   Bandwidth using Carson's rule:

   $$BW = 2 \left[\Delta f + f_m \right]$$

   $$BW = 2 \left[75 + 10 \right] = 170 \text{ kHz}$$

2. *The carrier frequency of a broadcast signal is* 50 MHz. *The maximum frequency deviation is* 60 kHz. *If the highest modulating frequency is limited to* 15 kHz, *then what is the approximate bandwidth of the modulating signal?*

   Given:
   Maximum frequency deviation = 60 kHz
   Modulating frequency = 15 kHz
   Answer:
   Bandwidth:

   $$BW = 2 \left[\Delta f + f_m \right]$$

   $$BW = 2 \left[60 + 15 \right] = 150 \text{ kHz}$$

3. *A carrier is frequency modulated with a sinusoidal signal of* 2 kHz *resulting in a maximum frequency deviation of* 5 kHz. *Find the bandwidth of modulated signal.*

   $$BW = 2 \left[\Delta f + f_m \right]$$

   $$BW = 2 \left[5 + 2 \right] = 14 \text{ kHz}$$

4. *A* 20 MHz *carrier is frequency modulated by a sinusoidal signal such that the maximum frequency deviation is* 100 kHz. *Find the modulation index and approximate bandwidth of FM signal, if the frequency of the modulating signal is* 100 kHz.

Given:
$$\Delta f_{\max} = 100 \, \text{kHz}; \, f_m = 100 \, \text{kHz}$$

Answer:
Modulation index $\beta_{\text{FM}} = \frac{\Delta f}{f_m} = 100/100 = 1$
Bandwidth:
$$BW = 2 \left[\Delta f + f_m \right]$$
$$BW = 2 \left[100 \, \text{kHz} + 100 \, \text{kHz} \right] = 400 \, \text{kHz}$$

5. *A 107.6 MHz carrier is frequency modulated by a 7 kHz sine wave. The resultant FM signal has a frequency deviation of 50 kHz. Find the modulation index of the FM wave.*

Given:
$$f_m = 7 \, \text{kHz}; \Delta f_{\max} = 50 \, \text{kHz}$$

Answer:
$$\beta_{\text{FM}} = \frac{\Delta f}{f_m} = \frac{50}{7} = 7.14$$

6. *Consider an angle modulated signal* $x_c(t) = 10 \cos(\omega_c t + 3 \sin \omega_m t)$. *Consider FM modulation technique and* $f_m = 1 \, \text{kHz}$, *calculate modulation index and bandwidth.*

Given:
Modulating signal frequency, $f_m = 1 \, \text{kHz}$ and
FM signal is, $x_c(t) = 10 \cos(\omega_c t + 3 \sin \omega_m t)$
Answer:
Compare given expression with generalized expression of FM signal.
Now generalized expression is

$$x(t) = A \cos \left[\omega_c t + \beta \sin \omega_m t \right]$$

Comparing this equation, it can be found, modulation index $\beta = 3$
Bandwidth:
$$BW = 2 \left[\Delta f + f_m \right]$$

Now Δf is unknown, but we know the relation

$$\beta_{\text{FM}} = \frac{\Delta f}{f_m}$$
$$\Delta f = \beta \times f_m = 3 \times 1 \, \text{kHz} = 3 \, \text{kHz}$$
$$BW = 2 \left[\Delta f + f_m \right]$$
$$BW = 2 \left[3 \, \text{kHz} + 1 \, \text{kHz} \right] = 8 \, \text{kHz}$$

7. *Consider an angle-modulated signal* $x_c(t) = 10\cos(\omega_c t + 3\sin\omega_m t)$. *Consider FM modulation technique and* $f_m = 1\,\text{kHz}$, *calculate modulation index and bandwidth, when a)* f_m *is doubled b)* f_m *is decreased by half.*

Given:
Modulating signal frequency, $f_m = 1\,\text{kHz}$ and
FM signal is $x_c(t) = 10\cos(\omega_c t + 3\sin\omega_m t)$
Answer:
Compare the given expression with the generalized expression of FM signal.
Now the generalized expression is

$$x(t) = A\cos[\omega_c t + \beta\sin\omega_m t]$$

Comparing this equation, we found, modulation index $\beta = 3$
Bandwidth:
$$BW = 2[\Delta f + f_m]$$

Now Δf is unknown, but we know the relation

$$\beta_{\text{FM}} = \frac{\Delta f}{f_m}$$

$$\Delta f = \beta \times f_m = 3 \times 1\,\text{kHz} = 3\,\text{kHz}$$

a) Bandwidth of the system, when f_m is doubled

$$BW = 2[3\,\text{kHz} + 2\,\text{kHz}] = 10\,\text{kHz}$$

b) Bandwidth of the system, when f_m is decreased by half

$$BW = 2\left[3\,\text{kHz} + \frac{1}{2}\text{kHz}\right] = 7\,\text{kHz}$$

8. *Determine the bandwidth occupied by a sinusoidal frequency carrier for which the modulation index is 2.4 and the maximum frequency deviation is 15 kHz.*

Given:
$$\beta = 2.4; \quad \Delta f = 15\,\text{kHz}$$

Answer:
Bandwidth:
$$BW = 2[\Delta f + f_m]$$

Here, unknown parameter is f_m;
We know that,

$$\beta_{\text{FM}} = \frac{\Delta f}{f_m}$$

Therefore,

$$f_m = \frac{\Delta f}{\beta} = \frac{15\,\text{kHz}}{2.4} = 6.25\,\text{kHz}$$

$$\text{BW} = 2\left[\Delta f + f_m\right] = 2\left[15\,\text{kHz} + 6.25\,\text{kHz}\right] = 42.5\,\text{kHz}$$

9. *Evaluate the thermal noise voltage developed across a resistor of* 700 Ω. *The bandwidth of the measuring instruments is* 7 *MHz, and the ambient temperature is* 27°C.

   Given: B = 7 MHz, R = 700Ω, Temperature (Kelvin) = $27°C + 273 = 300K$

   Answer:
   RMS Value of Thermal Noise voltage
   $V_N = \sqrt{4\text{KTRB}}$, Here K is Blotzman's constant

   $$= \sqrt{4 \times 1.38 \times 10^{-23} \times 300 \times 700 \times 7 \times 10^6}$$

   $$= \sqrt{81.1 \times 10^{-12}}$$

   $$= 9 \times 10^{-6} = 9\,\mu\text{V}.$$

10. *An amplifier operating over the frequency range from* 18 *to* 20 *MHz has a* 10 kΩ *input resistor. What is the rms noise voltage at the input to this amplifier if the ambient temperature is* 27°C.

    Given: Bandwidth (B) = 20MHz–18MHz = 2MHz.
    R = 10 kΩ, T (in Kelvin) = $27°C + 273 = 300$ K
    Answer:
    $$V_N = \sqrt{4\,\text{KTRB}}$$

    $$= \sqrt{4 \times 1.38 \times 10^{-23} \times 300 \times 10 \times 10^3 \times 2 \times 10^6}$$

    $$\approx 18.2\,\mu\text{V}$$

11. *For an electronics device operating at a temperature of* 17°C *with a bandwidth of* 10 *kHz. Find the thermal noise power.*

    Given: T (Kelvin) = $17°C + 273 = 290K$, B = 10 kHz.
    Solution: Thermal Noise power = KTB

    $$= \left(1.38 \times 10^{-23}\right) \times (290) \times \left(10 \times 10^3\right)$$

    $$= 4 \times 10^{-17}W$$

2.5 **Points to Remember**

- Types of analog modulation: AM, FM, and PM.
- Modulation: Changing the characteristic of carrier wave in accordance with the amplitude of modulating signal or message signal or information signal.
- Amplitude modulation: Changing the amplitude of the carrier in accordance with the amplitude of message signal.
- Modulation index: $m = \frac{E_m}{E_c}$
- Modulation index should be in the range of $0 < m < 1$.
  If $m \leq 1$, then the type of modulation is linear modulation.

- Total bandwidth of AM: $\text{BW} = 2 \times f_{m(\max)}$. Where $f_{m(\max)}$ is the maximum frequency of modulating signal.
- Percentage of modulation or depth of modulation $= m \times 100$.
- Efficiency of AM: $\eta = \frac{P_{\text{USB}} + P_{\text{LSB}}}{P_T} \times 100$.
- Relation between efficiency and modulation index:

$$\eta = \frac{m^2}{2 + m^2} \times 100.$$

- AM Total power calculation: $P_T = P_C \left(1 + (m^2/2)\right)$.
- AM Total current calculation: $I_T = I_C \sqrt{(1 + (m^2/2))}$.
- Types of AM: DSB-SC, SSB, and VSB.
- Total power in DSB-SC: $P_{T(\text{DSB-SC})} = P_C(m^2/2)$
- Carson's rule for bandwidth calculation: $\text{BW} = 2\left[\Delta f + f_m\right]$
- Total power requirement of FM: $P_T = \frac{E_c^2}{2R_L}$, where R_L-load resistance.
- Deviation ratio or percentage of modulation of FM:

$$D = \frac{\Delta f_{\text{mac}}}{f_{m(\max)}}$$

- Noise power: $N_O = \frac{P_N}{B_N}$.
- Noise temperature: $T_E = \frac{P_N}{KB_N}$.

PART II

Digital Communication

3

Digital Modulation

3.1 Introduction

As we know, a computer needs to communicate between various devices like mouse, keyboard, and printer or between various IC's. The information stored in these devices and computers are in binary format. (0s and 1 s). But the devices are connected only by means of electrical connection. Hence, we must require conversion of binary data into analog signal prior to transmission through the electrical wires. This type of conversion or mapping is called as Digital Modulation.

For example, Binary 0 by 0 V and Binary 1 by +5 V.

Types of Digital Modulation
Baseband technique.
Pass band technique.

Baseband Modulation
Low-frequency modulation is called the baseband modulation approach. Some of the examples are popular line coding approaches, like unipolar and polar formats.

Unipolar format
In this format, binary 1 is represented by some voltage and binary 0 by 0 V. For example, binary 1 by +5 V and binary 0 by 0 V. It is shown in Figure 3.1.

Polar format
In this format, a positive voltage is used to represent binary 1 and a negative voltage is used to represent binary 0. For example, binary 1 by +5 V and binary 0 by −5 V. It is shown in Figure 3.2.

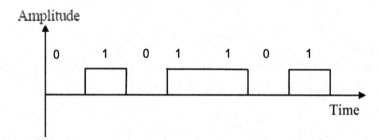

Figure 3.1   Unipolar format.

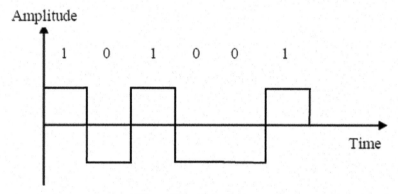

Figure 3.2   Polar format.

Pass band Modulation
It is a high-frequency modulation. In this approach, user binary data are multiplied by high-frequency (RF) carrier at the transmitter side.

Major types of pass band modulation are
Binary and M-ary modulation techniques.

Binary Pass band modulation
In binary modulation technique, bit 0 or 1 can be transmitted for every symbol time interval. The various binary modulation approaches are Amplitude shift keying (ASK), Phase shift keying (PSK), and Frequency Shift keying (FSK).

M-ary modulation
In this type, more than one bit can be transmitted for every symbol period. Some of the techniques are: QPSK (Quadrature Phase Shift Keying) – here, we transmit two binary digits at a time; M-PSK, M-FSK, and QAM (Quadrature

Amplitude modulation) – here, M represents the number of input levels. For example, in 8-PSK, 8 different input conditions can be transmitted or 3 binary digits at a time.

3.2 Binary Modulation Techniques

3.2.1 Amplitude Shift Keying

In ASK, binary digit 1 or symbol 1 is represented by some voltage and binary digit 0 or symbol 0 represented by 0 V. It is shown in Figure 3.3.

3.2.2 Mathematical Representation

ASK can be represented as follows,

Binary 1 or symbol 0 represented as $S_1(t)$ and binary 0 or symbol 0 represented as $S_2(t) \cdot S_1(t)$ and $S_2(t)$ can be written as

$$S_1(t) = \sqrt{\frac{2E_b}{T_b}} \cos 2\pi f_c t \quad 0 \leq t \leq T_b \quad \text{For Symbol 1 or Binary digit 1}$$

$$S_2(t) = 0 \quad 0 \leq t \leq T_b \quad \text{For Symbol 0 or Binary digit 0} \tag{3.1}$$

where E_b is signal energy per bit, T_b is the bit duration and f_c is carrier the frequency.

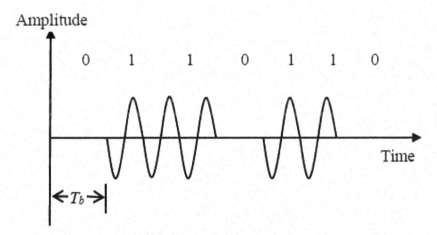

Figure 3.3 ASK waveform.

Here, orthonormal basis signal (or simply carrier signal) is

$$\varphi(t) = \sqrt{\frac{2}{T_b}} \cos 2\pi f_c t \quad 0 \le t \le T_b. \tag{3.2}$$

Now, we can represent $S_1(t)$ and $S_2(t)$ in terms of orthonormal basis function by

$$S_1(t) = \sqrt{E_b}\varphi(t) \quad 0 \le t \le T_b \quad \text{For Symbol 1 or Binary digit 1}$$

$$S_2(t) = 0 \quad 0 \le t \le T_b \quad \text{For Symbol 0 or Binary digit 0} \tag{3.3}$$

3.2.3 Signal Space Representation or Constellation Diagram

Signal space diagram helps us to define the amplitude and phase of the signal in a pictorial way. Figure 3.4 shows the constellation diagram of ASK modulation.

3.2.4 Probability of Error

Error probability of the modulation technique depends on the distance between two signal points, called "Euclidean distance".

Now the decision rules for detecting symbol $S_1(t)$ and $S_2(t)$ are as follows:

1. The set of received message points closest to the region R1 are termed as Symbol 1 or $S_1(t)$.

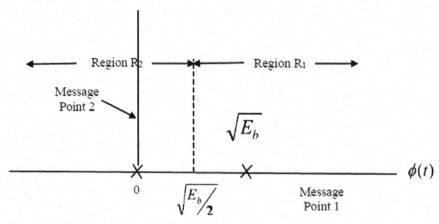

Figure 3.4   Signal space diagram-ASK.

2. The set of received message points closest to the region R2 are termed as Symbol 0 or $S_2(t)$.

Hence, probability of the error equation for ASK is

$$P_e = Q\left(\sqrt{\frac{E_b}{N_0}}\right) \tag{3.4}$$

In ASK, information is conveyed through the amplitude of the signal. Hence, ASK can easily suffer due to noise. Hence, the probabilty of error performance in ASK is poor compared with other schemes.

3.2.5 ASK Generation

Generation of a binary ASK signal requires representation of the input binary sequence in unipolar format with symbol 1 and 0 by $+\sqrt{E_b}$ and 0V, respectively. The resulting binary wave (unipolar format) and sinusoidal carrier (Basis signal) $\varphi(t) = \sqrt{2/T_b}\cos 2\pi f_c t$ are applied to a product modulator. The desired ASK signals are obtained at the modulator output. The circuit is shown in Figure 3.5.

3.2.6 ASK Reception or Demodulation

The noisy ASK signal is received at the receiver side. Detection of binary 1's and 0's from noisy ASK signal is done by applying $x(t)$ to the product modulator. The other input for the product modulator is a locally generated carrier signal $\varphi(t) = \sqrt{2/T_b}\cos 2\pi f_c t$. Now, this multiplication is sent to the correlator integrator. The basic principle for correlator detector is orthonormal basis operation.

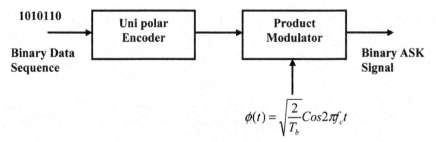

Figure 3.5   Generation of ASK signal.

Orthonormal basis operation is

$$\int_0^{T_b} \varphi_i(t)\,\varphi_j(t) = 1 \quad \text{when } i = j \text{ and } 0 \text{ for } i \neq j$$

1. Assuming receipt of binary 1, $x(t)$ is $\sqrt{2E_bT_b}\cos 2\pi f_c t$ (or) $\sqrt{E_b}\phi(t)$ and the other input of the product modulator is $\varphi(t) = \sqrt{2/T_b}\cos 2\pi f_c t$. When $x(t)$ and $\phi(t)$ are integrated, the output is $\sqrt{E_b} \cdot \left\{\int \sqrt{\frac{2E_b}{T_b}}\cos(2\pi f_c t) \cdot \sqrt{\frac{2}{T_b}}\cos 2\pi f_c t = \sqrt{E_b}\right\}$ (since received signal carrier frequency and locally generated carrier signal both are equal in phase). The correlator output $x_1 = \sqrt{E_b}$ is compared with a threshold λ, where $\lambda = \sqrt{E_b/2}$. Therefore, $x_1 > \lambda$, receiver decides in favor of symbol 1.
2. Assuming receipt of binary 0, $x(t)$ is 0 and the other input of the product modulator is $\varphi(t) = \sqrt{2/T_b}\cos 2\pi f_c t$. When $x(t)$ and $\phi(t)$ are integrated, the output is 0, since $\left\{\int 0 \times \sqrt{2/T_b}\cos 2\pi f_c t = 0\right\}$ The correlator output $x_1 = 0$ is compared with a threshold λ, where $\lambda = \sqrt{E_b/2}$. Therefore, $x_1 < \lambda$, receiver decides in favor of symbol 0. The receiver circuit is shown in Figure 3.6.

Bandwidth Requirement of ASK
ASK requires bandwidth, which is twice that of the modulating signal $S(t)$.

$$\text{BW} = 2 \times f_{\max} \tag{3.5}$$

where $f_{\max}$ is the maximum frequency of modulating signal $S(t)$.

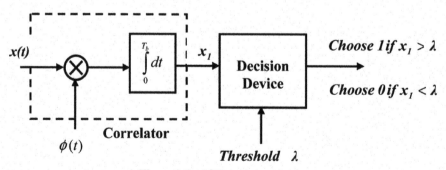

Figure 3.6   ASK detection circuit.

3.3 Phase Shift Keying

In PSK, binary digit 1 or symbol 1 is represented by some phase angle (for example, $0°$) and binary digit 0 or symbol 0 represented another phase angle (for example $180°$). It is shown in Figure 3.7.

3.3.1 Mathematical Representation

PSK can be represented as follows:

Binary 1 or symbol 1 is represented as $S_1(t)$ and binary 0 or symbol 0 is represented as $S_2(t) \cdot S_1(t)$ and $S_2(t)$ can be written as

$$S_1(t) = \sqrt{\frac{2E_b}{T_b}} \cos 2\pi f_c t \quad 0 \le t \le T_b \quad \text{For Symbol 1 or Binary digit 1}$$

$$S_2(t) = -\sqrt{\frac{2E_b}{T_b}} \cos 2\pi f_c t \quad 0 \le t \le T_b \quad \text{For Symbol 0 or Binary digit 0,}$$

$$(3.6)$$

where E_b is the signal energy per bit, T_b is the bit duration, and f_c is the carrier frequency.

Here, orthonormal basis signal (or simply carrier signal) is

$$\varphi(t) = \sqrt{\frac{2}{T_b}} \cos 2\pi f_c t \quad 0 \le t \le T_b \qquad (3.7)$$

Now $S_1(t)$ and $S_2(t)$ can be represented in terms of orthonormal basis function by

$$S_1(t) = \sqrt{E_b}\varphi(t) \quad 0 \le t \le T_b \quad \text{For Symbol 1 or Binary digit 1}$$

$$S_2(t) = -\sqrt{E_b}\varphi(t) \quad 0 \le t \le T_b \quad \text{For Symbol 0 or Binary digit 0}$$

$$(3.8)$$

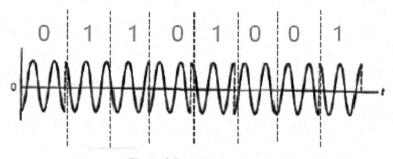

Figure 3.7 PSK waveform.

3.3.2 Signal Space Representation of PSK

The signal space diagram helps in pictorial representation of the amplitude and phase of the signal. Figure 3.8 shows the constellation diagram of PSK modulation.

In BPSK, signals $S_1(t)$ and $S_2(t)$ are differentiated with the help of 180° phase angle. Also $S_1(t) = \sqrt{E_b}$ and $S_2(t) = -\sqrt{E_b}$. Signals of this type are called Antipodal signal.

3.3.3 Probability of Error

Error probability of the modulation technique depends on the distance between two signal points. The Euclidean distance between the two points $S_1(t)$ and $S_2(t)$ is $2\sqrt{E_b}$.

Now, the decision rules for detecting symbols $S_1(t)$ and $S_2(t)$ are as follows:

1. The set of received message points closest to the region R1 is termed as Symbol 1 or $S_1(t)$.
2. The set of received message points closest to the region R2 is termed as Symbol 0 or $S_2(t)$.

Hence, the average probability of symbol error equation for PSK is

$$P_e = \frac{1}{2}erfc\left(\sqrt{\frac{E_b}{N_0}}\right) \qquad (3.9)$$

where $erfc\left(\sqrt{E_b/N_0}\right)$ is the complementary error function.

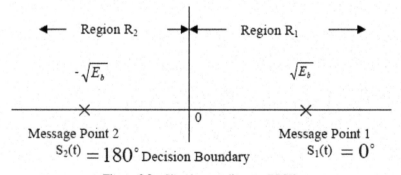

Figure 3.8   Signal space diagram-BPSK.

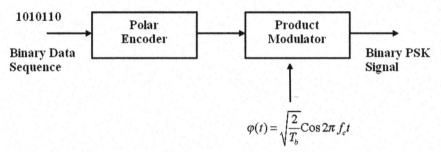

Figure 3.9 Generation of PSK signal.

3.3.4 PSK Generation

Generation of a binary PSK signal requires representation of the input binary sequence in *polar format* with symbols 1 and 0 by $+\sqrt{E_b}$ and $-\sqrt{E_b}$, respectively. The resulting binary wave (polar format) and sinusoidal carrier (Basis signal) $\varphi(t) = \sqrt{2/T_b} \cos 2\pi f_c t$ are applied to the product modulator. The desired PSK signals are obtained at the modulator output. The circuit is shown in Figure 3.9.

3.3.5 PSK Reception

The noisy PSK signal $x(t)$ is received at the receiver side. Detection of the binary 1's and 0's from noisy PSK signal is done by applying $x(t)$ to the product modulator. The other input for the product modulator is a locally generated carrier signal $\varphi(t) = \sqrt{2/T_b} \cos 2\pi f_c t$. Now, this multiplication is sent to a correlator integrator. The basic principle for correlator detector is the orthonormal basis operation.

Orthonormal basis operation is

$\int_0^{T_b} \phi_i(t) \phi_j(t) = 1$ when $i = j$ and 0 for $i \neq j$

1. Assuming receipt of binary 1, the received signal $x(t)$ is $\sqrt{2E_b/T_b}$ $\cos 2\pi f_c t$ (or) $\sqrt{E_b} \phi(t)$ and other input of the product modulator is $\varphi(t) = \sqrt{2/T_b} \cos 2\pi f_c t$, when $x(t)$ and $\phi(t)$ are multiplied and integrated for one symbol period, the output produced is $\sqrt{E_b}$ · $\left\{ \int \sqrt{\frac{2E_b}{T_b}} \cos(2\pi f_c t) \cdot \sqrt{\frac{2}{T_b}} \cos 2\pi f_c t = \sqrt{E_b} \right\}$ (since received signal carrier frequency and locally generated carrier signal both are equal in phase). The correlator output $x_1 = \sqrt{E_b}$ is then compared with a

threshold λ, where $\lambda = 0$. Therefore, $x_1 > \lambda$, receiver decides in favor of symbol 1.

2. Assuming receipt of binary 0, the received signal $x(t)$ is $-\sqrt{2E_b/T_b}$ $\cos 2\pi f_c t$ (or) $-\sqrt{E_b}\phi(t)$ and the other input of the product modulator is $\varphi(t) = \sqrt{2/T_b}\cos 2\pi f_c t$, when $x(t)$ and $\phi(t)$ are integrated, output produced is $-\sqrt{E_b} \cdot \left\{ \int (-) \sqrt{\frac{2E_b}{T_b}} \cos(2\pi f_c t) \cdot \sqrt{\frac{2}{T_b}} \cos 2\pi f_c t = -\sqrt{E_b} \right\}$ Then, the correlator output $x_1 = -\sqrt{E_b}$ is compared with a threshold λ, where $\lambda = 0$. Therefore, $x_1 < \lambda$, receiver decides in favors of symbol 0.

The receiver circuit is shown in Figure 3.10.

3.4 Frequency Shift Keying

In BFSK system symbol 1 or binary digit 1 and symbol 0 or binary digit 0 are differentiated with the help of different phase angles. For example, symbol 0 is represented by $0°$ and symbol 1 represented by another phase angle $180°$. In the same manner, in BFSK, symbols 1 and 0 are differentiated by different frequencies, i.e. symbol 0 represented by one frequency f_1 and symbol 1 represented by another frequency f_2. Hence, in FSK, with the help of frequencies, 0 and 1 are differentiated in FSK. Therefore, the modulation technique is called as Frequency shift keying. It is shown in Figure 3.11.

3.4.1 Mathematical Representation

FSK can be represented as follows,

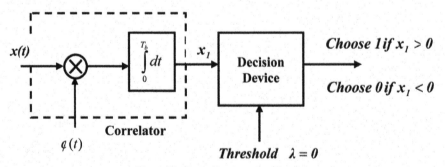

Figure 3.10 PSK detection circuit.

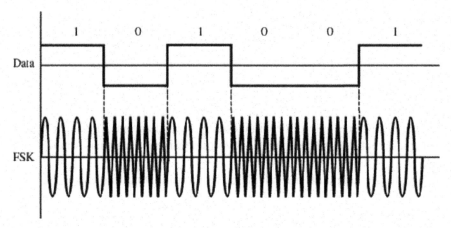

Figure 3.11 FSK waveform.

Binary 1 or symbol 0 is represented as $S_1(t)$ and binary 0 or symbol 0 is represented as $S_2(t) \cdot S_1(t)$ and $S_2(t)$ can be written as

$$S_1(t) = \sqrt{\frac{2E_b}{T_b}} \cos 2\pi f_1 t \quad 0 \le t \le T_b \quad \text{For Symbol 1 or Binary digit 1}$$

$$S_2(t) = \sqrt{\frac{2E_b}{T_b}} \cos 2\pi f_2 t \quad 0 \le t \le T_b \quad \text{For Symbol 0 or Binary digit 0,}$$

(3.10)

where E_b is the signal energy per bit, T_b is the bit duration, and f_1, f_2 are frequency figures for binary digits 1 and 0.

FSK can be represented in generalized form as follows:

$$S_i(t) = \sqrt{\frac{2E_b}{T_b}} \cos 2\pi f_i t \quad 0 \le t \le T_b,$$

(3.11)

where $f_i = n_c + i/T_b$ for some fixed integer n_c and $i = 1, 2$

Here, orthonormal basis signal (or simply carrier signal) is

$$\varphi_1(t) = \sqrt{\frac{2}{T_b}} \cos 2\pi f_1 t \quad 0 \le t \le T_b$$

(3.12)

$$\varphi_2(t) = \sqrt{\frac{2}{T_b}} \cos 2\pi f_2 t \quad 0 \le t \le T_b$$

(3.13)

3.4.2 Signal Space Representation of FSK

Signal space diagram helps in pictorial definition of the amplitude and the phase of the signal. Figure 3.12 shows the constellation diagram of FSK modulation.

Drawing the signal space diagram of BFSK needs finding the location of points in the constellation circles in the first place, since, previously discussed BASK and BPSK are single-dimensional signal space diagrams. (i.e., only one orthonormal function), whereas FSK is two-dimensional (two basis functions for FSK).

Hence, the signal points are calculated by the following formula:

$$S_{ij} = \int S_{ij}(t)\,\varphi_{ij}(t), \tag{3.14}$$

where $i, j = 1, 2$
When $i = 1$ and $j = 1$

$$S_{11} = \int S_{11}(t)\,\phi_{11}(t) = \sqrt{E_b}$$

(According to orthonormal basis property) $\tag{3.15}$

When $i = 1$ and $j = 2$

$$S_{12} = \int S_{12}(t)\,\phi_{12}(t) = 0 \tag{3.16}$$

When $i = 2$ and $j = 1$

$$S_{21} = \int S_{21}(t)\,\phi_{21}(t) = 0 \tag{3.17}$$

When $i = 2$ and $j = 2$

$$S_{22} = \int S_{22}(t)\,\varphi_{22}(t) = \sqrt{E_b}. \tag{3.18}$$

Therefore, the signal points are

$$s_1 = \left[\sqrt{E_b}, 0\right] \quad \text{and} \quad s_2 = \left[0, \sqrt{E_b}\right]. \tag{3.19}$$

Now the signal points can be plotted in the constellation diagram, the constellation diagram of BFSK can be obtained. This is shown in Figure 3.12.

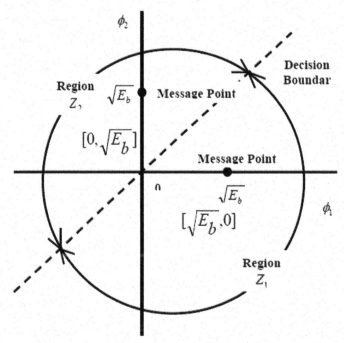

Figure 3.12 Signal space diagram of FSK.

3.4.3 Probability of Error

Error probability of the modulation technique depends on the distance between two signal points. The distance between the two points $S_1(t)$ and $S_2(t)$ is $\sqrt{2E_b}$ called the Euclidean distance.

Now the decision rules for detecting symbols $S_1(t)$ and $S_2(t)$ are as follows:

1. The set of received message points closest to the region Z_1 is termed as Symbol 1 or $S_1(t)$.
2. The set of received message points closest to the region Z_2 is termed as Symbol 0 or $S_2(t)$.

Hence, the average probability of symbol error equation for FSK is

$$P_e = \frac{1}{2}\text{erfc}\left(\sqrt{\frac{E_b}{2N_0}}\right) \tag{3.20}$$

When the probability of error formula for PSK and FSK are compared, the need for twice the bit energy-to-noise ratio (E_b/N_0) may be found in order to maintain the same probability of error in FSK. The main reason being the distance between the two message points. When a larger distance between two signal points is maintained, the probability of error is smaller. The distance between the two message points in FSK system is $\sqrt{2E_b}$, whereas in PSK system the distance is $2\sqrt{E_b}$.

3.4.4 FSK Generation

FSK generation circuit is shown in Figure 3.13. Input binary data sequences are sent to the on−off level encoder. The output of encoder is $\sqrt{E_b}$ volts for symbol 1 and 0 V for symbol 0. Assuming symbol 1, the upper channel is switched ON with a oscillator frequency of f_1. If symbol 0 is transmitted due to the inverter, the lower channel is switched ON with a oscillator frequency of f_2. These two frequencies are combined using an adder circuit. Now BFSK signal is available at the transmitter side.

3.4.5 FSK Reception

The detection circuit consists of two correlators. The inputs for the correlator circuits are received as noisy FSK signal $x(t)$, and another input is the basis function $\phi(t)$. For upper loop input, the correlators are $x(t)$ and $\phi_1(t)$ and input for lower correlator is $x(t)$ and $\phi_2(t)$. The correlator outputs are subtracted one from the other resulting in a random vector 1 or -1. Now the random vectors are compared with the threshold of zero volts. When $1 > 0$, the receiver decides in favor of symbol 1 and when $1 < 0$, the receiver decides in favor of symbol 0. It is shown in Figure 3.14.

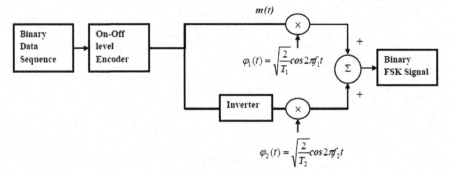

Figure 3.13 FSK generation.

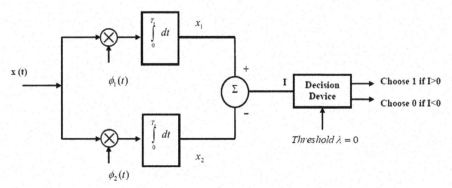

Figure 3.14 FSK reception.

3.5 Comparison of Binary Modulation Techniques

ASK versus PSK versus FSK

Probability of error is a performance tool for comparing modulation techniques. The probability of error equation for ASK, PSK, and FSK modulation techniques is known. We can also describe the probability of error in another manner, which can be found in the equation

$$P_e = Q\left(\sqrt{\frac{d_{12}^2}{2N_0}}\right).$$

Here, d_{12} represents the distance between two message points $S_1(t)$ and $S_2(t)$. When the distance between two message points is large, the probability of error is lower. The distance between the two points for various binary modulation techniques is shown in Table 3.1.

For example, let us take signal energy per bit E_b as 4 V.

Then for ASK-$d_{12} = \sqrt{4} = 2$;

For PSK-$d_{12} = 2\sqrt{4} = 2 \times 2 = 4$ and FSK-$d_{12} = \sqrt{2 \times 4} = \sqrt{8} = 2.82$.

It implies that PSK provides superior performance compared with other binary modulation techniques like ASK and FSK. For FSK, achievement of the same probability of error of PSK needs twice the signal energy per bit.

Table 3.1 Comparison of binary modulation techniques

ASK-	PSK-	FSK-
Distance between signal points: $d_{12} = \sqrt{E_b}$	Distance between signal points: $d_{12} = 2\sqrt{E_b}$	Distance between signal points: $d_{12} = \sqrt{2E_b}$

Figure 3.15 is the performance comparison chart for various modulation techniques.

- BPSK Vs BFSK: BPSK requires around 4 dB for achievement of a bit error rate (BER) of 10^{-2}, whereas BFSK requires around 7.5 dB. Hence, BPSK requires less SNR to achieve the same BER.
- At high SNR coherent PSK and QPSK provide the same performance.
- MSK has exactly the same error performance as QPSK. The main difference lies in MSK having memory at the receiver.

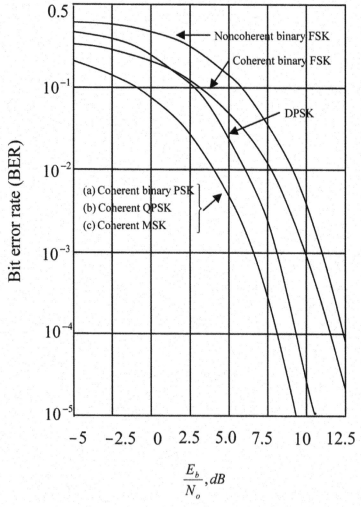

Figure 3.15   Comparison of BPSK versus BFSK versus MSK versus QPSK.

3.6 Quadrature Phase Shift Keying

In the binary PSK system, symbols 0 and 1 are differentiated by phase angles $0°$ and $180°$. In QPSK, two bits can be transmitted simultaneously (or we can transmit 2 bits per symbol interval). When the symbol size is two, transmission of four different informations is possible. The information may be 00 or 01 or 10 or 11. In QPSK, four informations are differentiated by four phase angles. Hence, the phase angle for different points of information are $\pi/4 = 45°, 3\pi/4 = 135°, 5\pi/4 = 225°$, and $7\pi/4 = 315°$.

3.6.1 Mathematical Representation

QPSK can be represented as follows,

$$S_i(t) = \sqrt{\frac{2E}{T}} \cos\left[2\pi f_c t + (2i-1)\frac{\pi}{4}\right] \quad 0 \leq t \leq T$$
$$\qquad\qquad 0 \qquad\qquad\qquad\qquad \text{elsewhere} \tag{3.21}$$

where $i = 1, 2, 3, 4$ $E-$symbol energy which is $E = 2E_b$, E_b is the bit energy, and T is the symbol duration.

Hence, when $i = 1$, signal representation for symbol 1 or data $1 = 00$

$$S_1(t) = \sqrt{\frac{2E}{T}} \cos\left[2\pi f_c t + \frac{\pi}{4}\right] \tag{3.22}$$

at $i = 2$; signal representation for symbol 2 or data $2 = 01$

$$S_2(t) = \sqrt{\frac{2E}{T}} \cos\left[2\pi f_c t + 3 \times \frac{\pi}{4}\right] \tag{3.23}$$

at $i = 3$; signal representation for symbol 3 or data $3 = 10$

$$S_3(t) = \sqrt{\frac{2E}{T}} \cos\left[2\pi f_c t + 5 \times \frac{\pi}{4}\right] \tag{3.24}$$

and at $i = 4$; signal representation for symbol 4 or data $4 = 11$

$$S_4(t) = \sqrt{\frac{2E}{T}} \cos\left[2\pi f_c t + 7 \times \frac{\pi}{4}\right]. \tag{3.25}$$

These are the mathematical representations of QPSK at various symbols.

Finding of basis function of QPSK is the next step, which needs expansion of Equation (1.1) by cos[A + B] formula. (cos $[A + B]$ = cos A cos B − sin A sin B)

$$S_i(t) = \sqrt{\frac{2E}{T}} \cos(2\pi f_c t) \cos\left[(2i-1)\frac{\pi}{4}\right]$$
$$0$$
$$-\sqrt{\frac{2E}{T}} \sin(2\pi f_c t) \sin\left[(2i-1)\frac{\pi}{4}\right] \quad 0 \le t \le T \tag{3.26}$$
$$\text{elsewhere}$$

Equation 3.26 shows that there are two basis functions of QPSK, and they are

$$\varphi_1(t) = \sqrt{\frac{2}{T_b}} \cos 2\pi f_c t \quad 0 \le t \le T \tag{3.27}$$

and

$$\varphi_2(t) = \sqrt{\frac{2}{T_b}} \sin 2\pi f_c t \quad 0 \le t \le T. \tag{3.28}$$

Now the four message points and the associated signal vector are defined by

$$S_i = \begin{bmatrix} \sqrt{E} \cos(2i-1)\frac{\pi}{4} \\ -\sqrt{E} \sin(2i-1)\frac{\pi}{4} \end{bmatrix} \quad \text{where } i = 1, 2, 3, 4. \tag{3.29}$$

The elements of the signal vector S_i are s_{i1} and s_{i2}. When the value of i is subtracted in the above equation, the values of signal vector may be found.

For example, if $i = 1$, phase angle $= \pi/4$; input dibit 00;

$$S_1 = \begin{bmatrix} +\sqrt{\frac{E}{2}} \\ -\sqrt{\frac{E}{2}} \end{bmatrix} \tag{3.30}$$

if $i = 2$; phase angle $= 3 \times (\pi/4)$; input dibit $= 01$;

$$S_2 = \begin{bmatrix} -\sqrt{\frac{E}{2}} \\ -\sqrt{\frac{E}{2}} \end{bmatrix} \tag{3.31}$$

if $i = 3$; phase angle $= 5 \times (\pi/4)$; input dibit $= 10$;

$$S_3 = \begin{bmatrix} -\sqrt{\frac{E}{2}} \\ +\sqrt{\frac{E}{2}} \end{bmatrix} \tag{3.32}$$

and if $i = 4$; phase angle $= 7 \times (\pi/4)$; input dibit $= 11$;

$$S_4 = \begin{bmatrix} +\sqrt{\frac{E}{2}} \\ +\sqrt{\frac{E}{2}} \end{bmatrix}. \tag{3.33}$$

The signal space diagram can be drawn on the basics of S_1, S_2, S_3, and S_4. Signal space diagram is shown in figure.

3.6.2 Signal Space Representation

Signal space diagram helps us to have a pictorial definition of the amplitude and the phase of the signal in a pictorial way. Figure 3.16 is the constellation diagram of QPSK modulation.

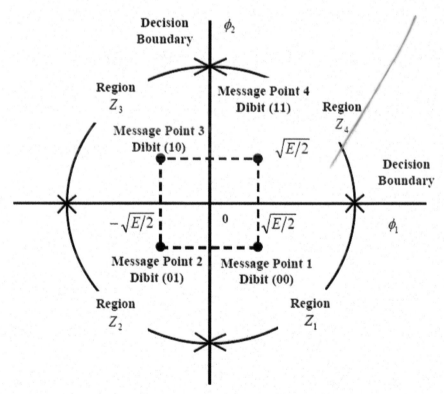

Figure 3.16   Signal space representation of QPSK.

3.6.3 Probability of Error

Error probability of the modulation technique depends on the distance between two signal points. The decision rules for detecting symbols $S_1(t)$, $S_2(t)$, $S_3(t)$, and $S_4(t)$ are as follows:

1. The set of received message points closest to the region Z1 is termed as Symbol 1 = 00 or $S_1(t)$.
2. The set of received message points closest to the region Z2 is termed as Symbol 2 = 01 or $S_2(t)$.
3. The set of received message points closest to the region Z3 is termed as Symbol 3 = 10 or $S_3(t)$.
4. The set of received message points closest to the region Z4 is termed as Symbol 4 = 11 or $S_4(t)$.

Hence, the average probability of symbol error equation for QPSK is

$$P_e = \text{erfc}\left(\sqrt{\frac{E_b}{N_0}}\right) \tag{3.34}$$

3.6.4 QPSK Generation

Figure 3.17 is a block diagram of QPSK transmitter. Generation of a QPSK signal requires representation of the input binary sequence in *polar format* with symbols 1 and 0 by $+\sqrt{E_b}$ and $-\sqrt{E_b}$, respectively. This binary wave is divided by means of demultiplexer. The purpose of the demultiplexer is to separate odd and even bits. The two binary waves $a_o(t)$ and $a_e(t)$ are

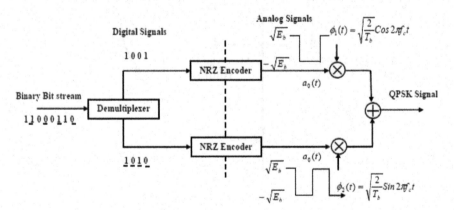

Figure 3.17   QPSK transmitter.

multiplied by basis signals $\phi_1(t)$ and $\phi_2(t)$. The result is a pair of BPSK signals and finally two binary PSK signals are added to produce the QPSK signal.

3.6.5 QPSK Reception

The QPSK receiver consists of a pair of correlators with a common input $x(t)$ and locally generated basis signals $\phi_1(t)$ and $\phi_2(t)$. The correlator outputs are x_1 and x_2, and they are compared with the threshold $\lambda = 0$.

Decision rule is,

When $x_1 > 0$, the decision is in favor of symbol 1, otherwise the decision is in favor of symbol 0. Similarly, when $x_2 > 0$ the decision is in favor of symbol 1, otherwise the decision is in favor of symbol 0. Finally, the output of decision circuits are combined with the help of a multiplexer (parallel to serial conversion) to produce the QPSK signal. It is shown in Figure 3.18.

3.7 Minimum Shift Keying

Why MSK? or Problem for QPSK
In QPSK, phase points of information are varied from $0°$ to $90°$ or $90°$ to $180°$ and so on. As a result, abrupt amplitude variations may be seen in the output waveform. Therefore, the bandwidth requirement of QPSK is more. MSK is introduced here for solving this issue.

In MSK, the output waveform is continuous in phase and hence there is no abrupt transition in amplitude. Side lobes are also very small in MSK

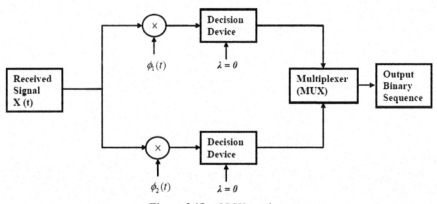

Figure 3.18 QPSK receiver.

waveform. Hence, band pass filter is not required for elimination of inter channel interference.

Therefore, MSK is a spectrally efficient modulation approach and finds application in mobile radio communication systems.

3.7.1 Mathematical Representation

MSK is a special type of continuous-phase frequency shift keying (CPFSK) system.

The transmitted signal is given by

$$S(t) = \begin{cases} \sqrt{\frac{2E_b}{T_b}} \cos\left[2\pi f_1 t + \theta(0)\right] & \text{For symbol 1} \\ \sqrt{\frac{2E_b}{T_b}} \cos\left[2\pi f_2 t + \theta(0)\right] & \text{For symbol 0} \end{cases} \tag{3.35}$$

where E_b is the signal energy per bit, T_b is the bit duration, and $\theta(0)$ is the phase of MSK signal at time $t = 0$.

CPFSK signal $S(t)$ can be represented in terms of an angle-modulated signal as follows:

$$S(t) = \sqrt{\frac{2E_b}{T_b}} \cos\left[2\pi f_c t + \theta(t)\right], \tag{3.36}$$

where phase $\theta(t)$ is,

$$\theta(t) = \theta(0) \pm \frac{\pi h}{T_b} t \quad 0 \leq t \leq T_b. \tag{3.37}$$

Phase of MSK $\theta(t)$ increases or decreases linearly with time during each bit duration of T_b seconds.

The transmitted frequencies f_1 and f_2 are given by,

$$f_1 = f_c + \frac{h}{2T_b} \tag{3.38}$$

and

$$f_2 = f_c - \frac{h}{2T_b}. \tag{3.39}$$

Carrier frequency $f_c = 1/2(f_1 + f_2)$ is the centre of the two signal frequencies.

The parameter h can be related to signal frequencies f_1 and f_2, and the relationship is given by subtracting Equation (3.38) from Equation (3.39),

$$h = T_b(f_1 - f_2)$$

Deviation ratio

The parameter h is called the deviation ratio. It is measured with respect to the bit rate $1/T_b$.

Now at $t = T_b$, the Equation (3.37) can be written as

$$\theta(T_b) = \theta(0) \pm \frac{\pi h}{T_b}T_b \tag{3.40}$$

$$\theta(T_b) - \theta(0) = \pm \pi h \tag{3.41}$$

$$\theta(T_b) - \theta(0) = \begin{array}{ll} \pi h & \text{for symbol 1} \\ -\pi h & \text{for symbol 0} \end{array} \tag{3.42}$$

Sending symbol 1 increases the phase of CPFSK signal by πh radians and sending symbol 0 decreases the phase of CPFSK signal by $-\pi h$ radians. When deviation ratio $h = 1/2$ is considered, the phase takes only two values. It may be $+\pi/2$ for symbol 1 or $-\pi/2$ for symbol 0.

Phase tree

The variation of phase $\theta(t)$ with time t follows a path consisting of a sequence of straight line, the slope of which represents frequency changes. This plot is known as a phase tree.

A binary sequence 1101000 with $\theta(0) = 0$ and $h = 1/2$ may be considered. The corresponding phase tree is shown in Figure 3.19.

Basis signal of MSK

CPFSK signal can be represented in terms of in-phase and quadrature components by applying $\cos(a + b) = \cos a \cos b - \sin a \sin b$; formulae to Equation (1.1)

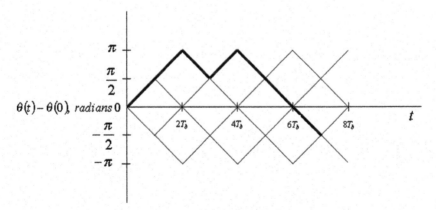

Figure 3.19 Phase tree for binary input sequence 1101000.

$$S(t) = \sqrt{\frac{2E_b}{T_b}} \cos[\theta(t)] \cos(2\pi f_c t) - \sqrt{\frac{2E_b}{T_b}} \sin[\theta(t)] \sin(2\pi f_c t). \tag{3.43}$$

The in phase component may be taken up for consideration

$$S_I(t) = \sqrt{\frac{2E_b}{T_b}} \cos[\theta(t)] \cos(2\pi f_c t). \tag{3.44}$$

Here, phase of CPFSK signal is subtracted in this equation, where

$$\theta(t) = \theta(0) \pm \frac{\pi h}{T_b} t \tag{3.45}$$

and put $h = \frac{1}{2}$ in the interval of $0 \le t \le T_b$

$$S_I(t) = \sqrt{\frac{2E_b}{T_b}} \cos\left[\theta(0) \pm \frac{\pi}{2T_b} t\right] \cos(2\pi f_c t) \tag{3.46}$$

$$S_I(t) = \sqrt{\frac{2E_b}{T_b}} \cos[\theta(0)] \cos\left[\frac{\pi}{2T_b} t\right] \cos(2\pi f_c t)$$

$$S_I(t) = \pm\sqrt{\frac{2E_b}{T_b}} \cos\left[\frac{\pi}{2T_b} t\right] \cos(2\pi f_c t) \tag{3.47}$$

Here $+$ sign corresponds to $\theta(0) = 0$ and the negative sign corresponds to $\theta(0) = \pi$.

Therefore, first basis signal $\phi_1(t)$ is

$$\varphi_1(t) = \sqrt{\frac{2}{T_b}} \cos\left[\frac{\pi}{2T_b} t\right] \cos(2\pi f_c t). \tag{3.48}$$

Therefore,

$$S_I(t) = \sqrt{E_b} \cos[\theta(0)] \phi_1(t) \tag{3.49}$$

Similarly, the quadrature component can also written in the interval of $0 \le t \le 2T_b$ by

$$\begin{aligned} S_Q(t) &= \sqrt{\frac{2E_b}{T_b}} \sin\left[\theta(T_b) \pm \frac{\pi}{2T_b} t\right] \sin(2\pi f_c t) \\ &= \sqrt{\frac{2E_b}{T_b}} \sin[\theta(T_b)] \sin\left[\frac{\pi}{2T_b} t\right] \sin(2\pi f_c t) \\ &= \pm\sqrt{\frac{2E_b}{T_b}} \sin\left[\frac{\pi}{2T_b} t\right] \sin(2\pi f_c t), \end{aligned} \tag{3.50}$$

where the plus sign corresponds to $\theta(T_b) = \pi/2$ and the negative sign corresponds to $\theta(T_b) = -\pi/2$.

Therefore, second basis signal $\phi_2(t)$ is

$$\phi_2(t) = \sqrt{\frac{2}{T_b}} \sin\left[\frac{\pi}{2T_b}t\right] \sin(2\pi f_c t) \tag{3.51}$$

$$S_Q(t) = -\sqrt{E_b} \sin[\theta(T_b)] \varphi_2(t). \tag{3.52}$$

Therefore,

$$S(t) = S_I(t)\varphi_1(t) + S_Q(t)\varphi_2(t). \tag{3.53}$$

The conclusion is that phase values of $\theta(0)$ and $\theta(T_b)$ can take one of the two possible values, that may be following combinations;

a. The phase $\theta(0) = 0$ and $\theta(T_b) = \pi/2$, corresponding to the transmission of symbol 1.
b. The phase $\theta(0) = \pi$ and $\theta(T_b) = \pi/2$, corresponding to the transmission of symbol 0.
c. The phase $\theta(0) = \pi$ and $\theta(T_b) = -\pi/2$, corresponding to the transmission of symbol 1.
d. The phase $\theta(0) = 0$ and $\theta(T_b) = -\pi/2$, corresponding to the transmission of symbol 0.

3.7.2 Signal Space Representation

Signal space diagram helps pictorial definition of the amplitude and the phase of the signal. Figure 3.20 shows the constellation diagram of MSK modulation.

When the signal space diagrams of QPSK and MSK are compared, both look similar. They differ by following points:

1. In QPSK, message points are expressed as signal energy per symbol E. In MSK, the same can be expressed as signal energy per bit E_b and $E_b = E/2$.
2. In QPSK, orthonormal basis signals $\phi_1(t)$ and $\phi_2(t)$ are a pair of quadrature carriers. In MSK, $\phi_1(t)$ and $\phi_2(t)$ are a pair of sinusoidally modulated quadrature carriers.

3.7.3 Probability of Error

The error probability of the modulation technique depends on the distance between two signal points. Now the decision rules for detecting message points 1, 2, 3, and 4 are as follows:

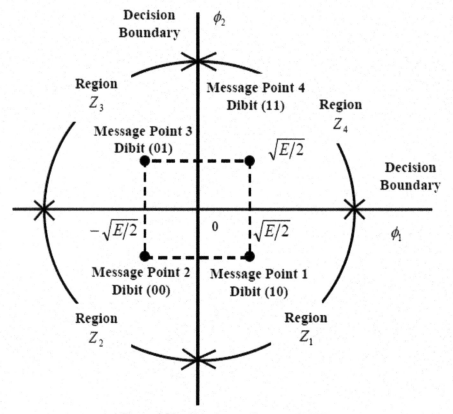

Figure 3.20   Signal space diagram of MSK.

1. The set of received message points closest to the region Z1 is termed as Symbol 1 = 10.
2. The set of received message points closest to the region Z2 is termed as Symbol 2 = 00.
3. The set of received message points closest to the region Z3 is termed as Symbol 3 = 01.
4. The set of received message points closest to the region Z4 is termed as Symbol 4 = 11.

Hence, the average probability of symbol error equation for QPSK is

$$P_e = \text{erfc}\left(\sqrt{\frac{E_b}{N_0}}\right) \tag{3.54}$$

3.7.4 MSK Generation

Two input sinusoidal waves, one at frequency $f_c = n_c/4T_b$, and the other at frequency of $1/4T_b$, where n_c is some fixed integer applied to the product modulator. The output of product modulator consists of two sinusoidal waves, each separated by narrow band filters. One filter is centered at f_1 and the other at f_2. The filter outputs are summed up to produce the orthonormal basis function $\phi_1(t)$ and $\phi_2(t)$. Finally, $\phi_1(t)$ and $\phi_2(t)$ are multiplied by two binary waves $m_1(t)$ and $m_2(t)$. The resultant signals are summed up to produce MSK waveform $S(t)$. It is shown in Figure 3.21.

3.7.5 MSK Demodulation

The received noisy MSK signals $x(t)$ are multiplied by locally generated orthonormal carriers $\phi_1(t)$ and $\phi_2(t)$. The outputs are integrated over the period of $2T_b$. The quadrature channel is delayed by T_b seconds with respect to in-phase channel.

In-phase and quadrature channel outputs are termed as $x1$ and $x2$ and they are compared with the threshold of zero, and estimates the phase of $\hat{\theta}(0)$ and $\hat{\theta}(T_b)$. Finally, these phase informations are interleaved so as to reconstruct the binary sequence at receiver side. It is shown in Figure 3.22.

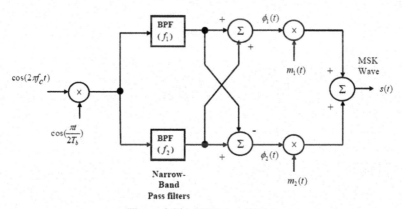

Figure 3.21   MSK transmitter.

Following Table 3.2 lists the comparison of QPSK and MSK schemes.

Table 3.2   Comparison between QPSK and MSK

S.No.	QPSK	MSK
1	QPSK signals have abrupt amplitude variations.	MSK signals have continuous phase variations.

(Continued)

Table 3.2 Continued

S.No.	QPSK	MSK
2	The main lobe of QPSK contains around 90% of signal energy.	The main lobe of MSK contains 99% of signal energy.
3	Side lobes are large. QPSK needs band pass filter in operation for avoiding inter-channel interference (ICI).	Side lobes are very small. Hence, MSK system is free from ICI and band pass filter.

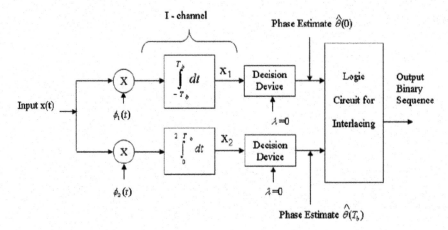

Figure 3.22   MSK receiver.

3.8 M-ary Modulation Techniques

In binary digital modulation techniques like BASK, BPSK, and BFSK, only one bit a time is transmitted. Here, one symbol = 1 bit, whereas in M-ary modulation technique, n-bit can be sent at a time. Here, one symbol = n number of bits. The variable 'n' can be related to 'M' by following formulae

$$M = 2^n \qquad (3.55)$$

This equation establishes the ability to transmit any one of M different symbols for every symbol period. For example, QPSK modulation technique may be considered. Here, two bits are transmitted for every symbol period. As per the 'M' ary formulae, it is possible to transmit $M = 2^2 = 4$; four different symbols in QPSK modulation technique.

The conclusion is that M represents the total number of symbols that can be transmitted from the modulation technique and the number of bits per symbol is defined by 'n'.

Example: 1) If the modulation technique is named as 8-PSK, then what it implies?

Answer: 8-PSK refers, $M = 8$, i.e. 8 different symbols can be transmitted from an 8-PSK modulation approach. Also, the relation between 'M' and 'n' can be found; the conclusion from is that, for every symbol period 8-PSK, modulation sends 3-bit per symbol. Since $8 = 2^n; n = 3$.

Example: 2) If the modulation technique is named as 16-PSK, then what it implies?

Answer: 16-PSK refers, $M = 16$, i.e. we can be able to transmit 16 different symbol and the number of bits per symbol are 4-bits. Since, according to the formulae $M = 2^n; 16 = 2^n; n = 4$.

Different types of M-ary modulation technique

- M-ary ASK
- M-ary-PSK
- M-ary-FSK
- QAM technique.

3.8.1 M-ASK

In binary ASK (BASK), only one bit can be transmitted for every symbol interval. It can be 0 or 1. This 0 and 1 can be differentiated with the help of two different amplitudes. For example, Binary 1 can be represented by +5 V and binary data 0 can be represented by 0 V. In M-ary ASK, M-different symbols are differentiated by M-different amplitudes.

For example, 4-ASK, $M = 4$, which means that four different symbols are differentiated by four different amplitudes and each symbol carries two bits at a time. Since $4 = 2^n; n = 2$.

Four symbols in detail may be;

symbol 1 = 00 represented by -3 V

symbol 2 = 01, represented by -1 V

symbol 3 = 11, represented by $+1$ V

and symbol 4 = 10 represented by $+3$ V

What is 8-ASK?

Answer: Here, $M = 8$. Therefore, the number of symbols transmitted by a particular modulation technique is 8 and each symbol carries 3 bits.

Eight different amplitude levels are needed for differentiating eight different symbols.

2-ASK, 4-ASK, and 8-ASK are signal space diagrams shown in Figure 3.23.

Mathematical representation of M-ASK

$$S_n(t) = A_n \cos(2\pi f_c t) \tag{3.56}$$

where $n = 1, 2, ...M$, and carrier frequency f_c is common for all the symbols.

3.9 M-PSK

In the binary PSK system, symbols 1 and 0 are differentiated by two phase angles. It can be $0°$ for symbol 1 and $180°$ for symbol 0. In M-ary PSK, M-different symbols are identified or differentiated by M-different phase angles.

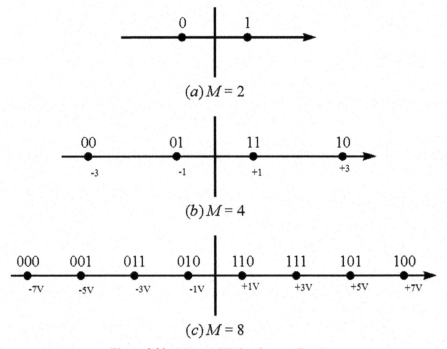

Figure 3.23   M-ary ASK signal space diagram.

For example, in the 8-PSK system, eight different symbols are differentiated by eight different phase angles and each symbol carries three bits.

Eight different phase angles can be $0°, 45°, 90°, 135°, 180°, 225°, 270°,$ and $315°$. Signal space diagram of 8-PSK is shown in Figure 3.24. Eight message points are equally spaced on a circle of radius $\sqrt{E}$ as illustrated in the figure.

3.9.1 Mathematical Representation

In M-ary PSK, the phase of the carrier takes any one of M possible values, namely $\theta_i = {}^{2i\pi}/_M$, where $i = 0, 1, \ldots\ldots M - 1$. The M-ary PSK signal can be mathematically represented as,

$$S_i\left(t\right) = \sqrt{\frac{2E}{T}} \cos\left(2\pi f_c t + \frac{2i\pi}{M}\right) \quad i = 0, 1, \ldots, M - 1 \qquad (3.57)$$

where E-is signal energy per symbol, T- symbol duration.

When the above equation is expanded by $\cos\left(a + b\right) = \cos a \,.\, \cos b - \sin a \,.\, \sin b$ formulae, two basis functions are found. They are

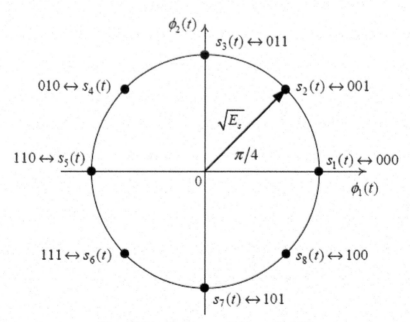

Figure 3.24   Signal space diagram of 8-PSK.

$$\varphi_1(t) = \sqrt{\frac{2}{T}} \cos(2\pi f_c t) \quad 0 \le t \le T \tag{3.58}$$

$$\varphi_2(t) = \sqrt{\frac{2}{T}} \sin(2\pi f_c t) \quad 0 \le t \le T \tag{3.59}$$

3.9.2 Receiver Circuit for M-PSK

The receiver circuit includes a pair of correlators. The output of correlators are r_1 and r_2. Then compute $(r_1 - s_{i1})^2 + (r_2 - s_{i2})^2$ and choose the smallest one. Figure 3.25 shows the decision region for symbols 1 and 2. The same decision rule can be applied to all other symbols. It is shown in Figure 3.26.

Probability of Error
The probability of symbol error is given by

$$P_e = \text{erfc}\left(\sqrt{\frac{2E}{N_0}} \sin\left(\frac{\pi}{2M}\right)\right) \quad M \ge 4 \tag{3.60}$$

3.10 M-QAM

In M-ary ASK, we can differentiate various symbol by different amplitude levels. Where as in M-ary PSK, we can identify various symbol by different phase angle.

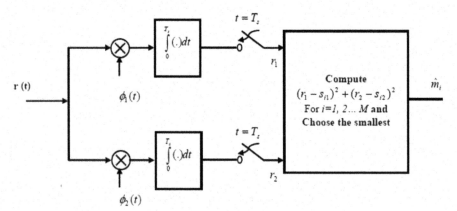

Figure 3.25   M-ary PSK receiver.

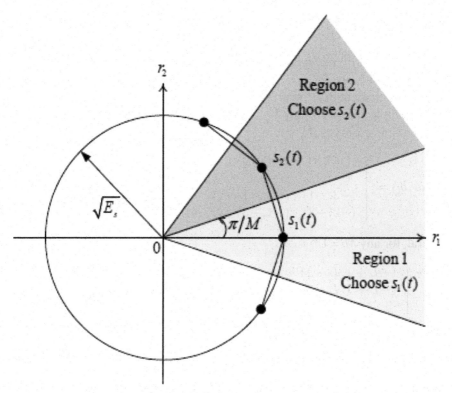

Figure 3.26   Decision region for M-ary PSK.

But in M-ary QAM, we can differentiate various symbols by combination of amplitude and phase angle. Hence, it is called as Quadrature Amplitude Modulation (QAM).

Mathematically, we can describe QAM technique by

$$S_i(t) = \sqrt{\frac{2E_0}{T}} a_i \cos(2\pi f_c t) + \sqrt{\frac{2E_0}{T}} b_i \sin(2\pi f_c t) \quad 0 \le t \le T \quad (3.61)$$

where E_0 is the energy of the signal with minimum amplitude and a_i and b_i are pair of independent integers.

Set of orthonormal basis signals are

$$\varphi_1(t) = \sqrt{\frac{2}{T}} \cos(2\pi f_c t) \quad 0 \le t \le T \tag{3.62}$$

$$\varphi_2(t) = \sqrt{\frac{2}{T}} \sin(2\pi f_c t) \quad 0 \le t \le T \tag{3.63}$$

Hence,

$$S_i(t) = \sqrt{E_0} a_i \phi_1(t) + \sqrt{E_0} b_i \phi_2(t) \tag{3.64}$$

The coordinates of ith message point depends on $\sqrt{E_0} a_i$ and $\sqrt{E_0} b_i$

where a_i and b_i are

(a_i, b_i) = element in L^2 matrix, where $L = \sqrt{M}$

For example, $M = 16$ may be considered. Then (a_i, b_i) are based on,
$\sqrt{E_0} a_i \phi_1(t) + \sqrt{E_0} b_i \phi_2(t)$

$$(a_i, b_i) = \begin{bmatrix} (-3,3) & (-1,3) & (1,3) & (3,3) \\ (-3,1) & (-1,1) & (1,1) & (3,1) \\ (-3,-1) & (-1,-1) & (1,-1) & (3,-1) \\ (-3,-3) & (-1,-3) & (1,-3) & (-3,-3) \end{bmatrix} \tag{3.65}$$

In general, for any $M = L^2$

$$(a_i, b_i) = \begin{bmatrix} (-L+1, L-1) & (-L+3, L-1) & \dots & (L-1, L-1) \\ (-L+1, L-3) & (-L+3, L-3) & \dots & (L-1, L-3) \\ \dots & \dots & \dots & \dots \\ (-L+1, -L+1) & (-L+3, -L+1) & \dots & (L-1, -L+1) \end{bmatrix}$$

$$\tag{3.66}$$

Signal Space diagram of 8-QAM, 16-QAM

The Figure 3.27 is the constellation diagram of 8-QAM, which can send 8 different symbols. Each symbol is indentified by a unique amplitude and phase angle. For example, in the above figure, there are four symbols inside the circle points defined by amplitude a_i and phase angles $0°, 90°, 180°,$ and $270°$, whereas other four symbol points outside the circle are defined by another amplitude b_i and phase angles $45°, 135°, 225°,$ and $315°$. Signal constellation for 16-QAM is shown in Figure 3.28.

Probability of Error

The average error probability of M-ary QAM is

$$P_e = 4\left(1 - \frac{1}{\sqrt{M}}\right) Q\left(\sqrt{\frac{2E_{\min}}{N_0}}\right) \tag{3.67}$$

Power spectrum and bandwidth efficiency of QAM system are equal to MPSK, but power efficiency of QAM is better than MPSK as it requires uniform energy for all the symbols.

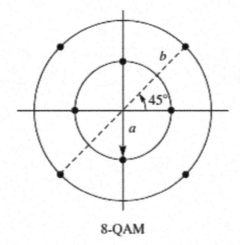

8-QAM

Figure 3.27 Signal constellation diagram of 8-QAM.

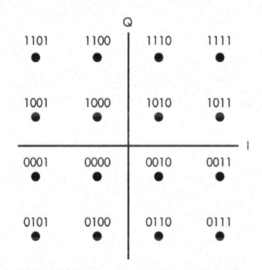

Figure 3.28 Signal constellation of 16-QAM.

3.10.1 M-ary QAM Transmitter

Figure 3.29 shows the M-ary QAM transmitter. Input binary sequences are given to serial-to-parallel converter, producing two parallel binary sequences. The 2-to-L converter generates polar L-level signals in response to the respective in-phase and quadrature channel inputs. Quadrature carrier multiplexing of the two polar L-level signals generate the M-ary QAM signal.

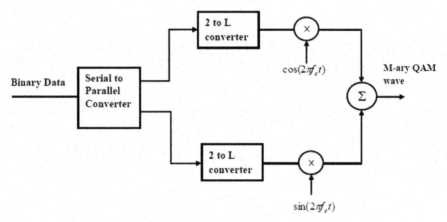

Figure 3.29   QAM transmitter.

3.10.2 M-ary QAM Receiver

Figure 3.30 shows the M-ary QAM receiver circuits. The received signals are multiplied by the basis signal and given to the correlator circuit.
The decision circuit is designed to enable comparison of L-level signals
against L-1 decision thresholds. The two detected binary sequences are
combined at parallel-to-serial converter in order to produce the original binary
sequence.

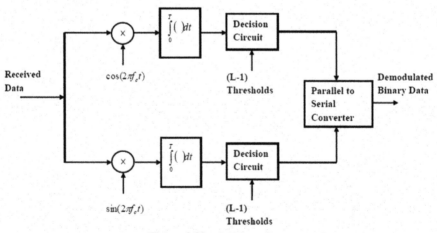

Figure 3.30   QAM receiver.

3.11 M-FSK

In M-ary FSK, M different symbols are differentiated by M different frequencies. We can write it as this mathematically

$$S_i(t) = \sqrt{\frac{2E}{T}} \cos\left[\frac{\pi}{T}(n_c + i)\,t\right] \quad 0 \le t \le T, \tag{3.68}$$

where $i = 1, 2, \ldots M$, carrier frequency $f_c = n_c/2T$ and n_c is the fixed integer.

The above equation can be written as

$$S_i(t) = \sqrt{\frac{2E_s}{T_s}} \cos\left(2\pi\left(f_c + \frac{i}{2T_s}\right)t\right). \tag{3.69}$$

It shows all M signals having equal energy, equal duration with adjacent sub carriers separated by $1/(2T)$ Hz, and the subcarriers are orthogonal to each other.

MFSK Detection

Receiver circuit has bank of M-correlators. Each correlator is tuned to one of M distinct carrier frequencies.

The average probability of error for M-ary FSK system is given by

$$P_e \le \frac{1}{2}(M-1)\,\mathrm{erfc}\left(\sqrt{\frac{E}{2N_0}}\right). \tag{3.70}$$

3.12 Comparison of Various M-ary Modulation Techniques

A comparison can be made of the various M-ary modulation techniques with the help of two important parameters, namely bandwidth and power. Efficient utilization of these parameters is very important in any communication system.

Bandwidth efficiency: It is defined as the ratio of data rate to channel bandwidth. It is measured in units of bits per second per hertz. It is also referred to as spectral efficiency. This can be given mathematically as

$$\rho = \frac{R_b}{B}\,\mathrm{bits/s/Hz}, \tag{3.71}$$

where R_b is the data rate and B the channel bandwidth.

M-ary PSK

Bandwidth efficiency of M-ary PSK:

$$\rho = \frac{\log_2 M}{2} \tag{3.72}$$

For different values of M, bandwidth efficiency ρ can be calculated. Details are given in Table 3.3.

The table shows improvement in bandwidth efficiency when the value of M is increased.

Power efficiency

Power efficiency refers to the efficient usage of power in modulation operation. The use of power should be to the minimum extent possible. Power efficiency figures are plotted in Table 3.4 for various values of M.

In this table, E_b/N_0 refers to bit error rate (BER), on the assumption BER of 10^{-6}. It shows $M = 4$ PSK requiring 10.5 W to achieve BER of 10^{-6}., whereas $M = 8$ PSK requires BER of 14 W. It implies, when the value of M is increased, power also increases. Hence, M-ary PSK proves to be poor in power efficiency and better in bandwidth efficiency.

M-ary FSK

In M-ary FSK, bandwidth efficiency is given by

$$\rho = \frac{2\log_2 M}{2} \tag{3.73}$$

Bandwidth efficiency and power efficiency figures for various levels of M are plotted in Table 3.5.

In M-ary FSK, bandwidth efficiency reduces, whereas power efficiency improves. For example, achievement of the BER of 10^{-6}, at $M = 8$ requires 9.3 W while $M = 64$ requires only 6.9 W. Hence, M-ary FSK is good for power efficiency and poor in bandwidth efficiency.

Table 3.3   Bandwidth efficiency of M-ary PSK system

M	2	4	8	16	32	64
$\rho_B = R_b/B_{null}$	0.5	1.0	1.5	2	2.5	3

Table 3.4   Power efficiency of M-ary PSK

M	2	4	8	16	32	64
E_b/N_0 (dB)(BER $= 10^{-6}$)	10.5	10.5	14.0	18.5	23.4	28.5

Table 3.5 Bandwidth and power efficiency of M-ary FSK

M	2	4	8	16	32	64
$\eta_B = R_b/B_{null}$	0.4	0.57	0.55	0.42	0.29	0.1K
$E_b N_0 (BER = 10^{-6})$	13.5	10.8	9.3	8.2	7.5	6.9

M-ary QAM

Bandwidth and power efficiency figures of M-ary QAM are plotted in table for various values of M. It can be shown in Table 3.6.

The conclusion is that, M-ary QAM and M-ary PSK have identical bandwidth and power efficiency. However, when $M > 4$, these two schemes have different signal constellations in signal space diagram. M-ary PSK is circular in nature, whereas M-ary QAM is rectangular. It shows that the distance between the two points is very small in M-ary PSK, when we increase the value of M compared with M-ary QAM. Hence M-ary QAM outperforms in terms of error performance.

3.13 Points to Remember

- ASK Waveform

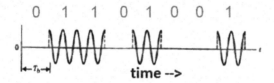

- PSK Waveform

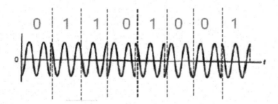

Table 3.6 Bandwidth and power efficiency of M-ary QAM

M	4	16	64	256	1024	4096
$\eta_B = R_b/B_{null}$	1	2	3	4	5	6
$E_b N_0 (BER = 10^{-6})$	10.5	15	18.4	24	28	33.5

- FSK Waveform

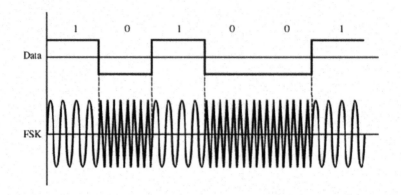

- Euclidean distance: The distance between two message points in signal space diagram is referred as Euclidean distance.

ASK-Euclidean distance:
$$d_{12} = \sqrt{E_b}$$

PSK-Euclidean distance:
$$d_{12} = 2\sqrt{E_b}$$

FSK-Euclidean distance:
$$d_{12} = \sqrt{2E_b}$$

- Probability of error

PSK-
$$P_e = \tfrac{1}{2}\text{erfc}\left(\sqrt{\tfrac{E_b}{N_0}}\right)$$

M-PSK
$$P_e = \text{erfc}\left(\sqrt{\tfrac{2E}{N_0}}\sin\left(\tfrac{\pi}{2M}\right)\right)$$
$$M \geq 4$$

MSK
$$P_e = \text{erfc}\left(\sqrt{\tfrac{E_b}{N_0}}\right)$$

FSK-
$$P_e = \tfrac{1}{2}\text{erfc}\left(\sqrt{\tfrac{E_b}{2N_0}}\right)$$

M-FSK
$$P_e \leq \tfrac{1}{2}(M-1)\,\text{erfc}\left(\sqrt{\tfrac{E}{2N_0}}\right)$$

QPSK-
$$P_e = \text{erfc}\left(\sqrt{\tfrac{E_b}{N_0}}\right)$$

M-QAM
$$P_e = 4\left(1 - \tfrac{1}{\sqrt{M}}\right)Q\left(\sqrt{\tfrac{2E_{\min}}{N_0}}\right)$$

PART III

Pulse and Data Communication

4

Pulse Modulation

4.1 Introduction

The process of changing the characteristics of pulse carrier in accordance with the modulating signal is called pulse modulation. The types of pulse modulations are: analog pulse modulation and digital pulse modulation. Classifications of analog pulse modulation:

The important characteristics of analog pulse carriers are amplitude, width, and position. The various types of analog pulse modulation techniques based on these characteristics are:

1. Pulse amplitude modulation (PAM)
2. Pulse width modulation (PWM)
3. Pulse position modulation (PPM)

Types of Digital Pulse modulation:

1. Pulse code modulation (PCM)
2. Delta modulation
3. Adaptive delta modulation

The major points of difference between analog modulation and pulse modulation are: in analog modulation technique, simple or complex sinusoidal signal is considered as a carrier signal, whereas in pulse modulation, carrier signals are periodic rectangular trains of pulse signals. This is shown in Figure 4.1.

4.2 Pulse Amplitude Modulation

The process of changing the amplitude of the pulse carrier signal in accordance with the modulating signal is called "Pulse Amplitude Modulation" (PAM) which is also referred to as *"sampling process"*.

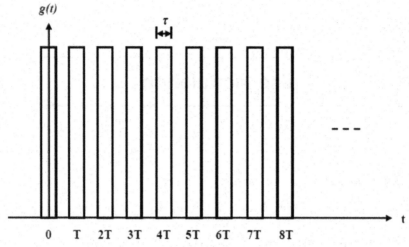

Figure 4.1 Carrier signal format in pulse modulation.

4.2.1 Generation of PAM Signals/Sampling Process

PAM signals can be generated using electronic switches. The inputs for the electronic switches are continuous-time-modulating signal and train of periodic pulses. When the switch is closed, the corresponding instant message signals are arrived at the output side. The output is zero when the switch is open. Continuous time signals are converted into discrete-time signal due to this process. The output signals are sampled signals. Figure 4.2 represents the sampler as a switch. Figure 4.3 shows the input waveform, carrier signal, and sampled output waveform of sampling operation.

Let $x(t)$ be the modulating signal (continuous)

$g(t)$ be the carrier (pulse) signal and $x_s(t)$ sampled signal.

Now $g(t)$ are periodic trains of pulses. Any periodic signal can be represented by Fourier series. Hence, Fourier series representation of carrier signal $g(t)$ is given by

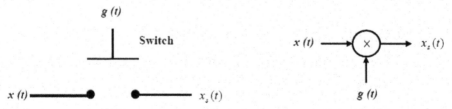

Figure 4.2 (a) Modeling samples as a switch. (b) Model of a sampler.

Sampling operation:
  (a) Input or message signal = $x(t)$

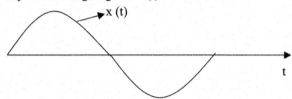

(b) Carrier signal = $g(t)$

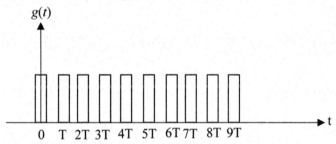

(c) Sampled signal = $x_s(t)$

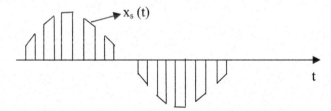

Figure 4.3   (a) Input signal (b) Carrier signal (c) Sampled signal.

$$g(t) = \sum_{n=-\infty}^{\infty} C_n e^{j2\pi n f_s t} \tag{4.1}$$

where C_n is a nth Fourier coefficient and it can be expressed as

$$C_n = \frac{1}{T} \int_{-\frac{f_s}{2}}^{\frac{f_s}{2}} g(t) e^{-j2\pi n f_s t} \tag{4.2}$$

f_s is the sampling frequency.

The output of a electronic switch or sampler circuit is given as

$$x_s(t) = x(t) g(t) \qquad (4.3)$$

This expression describes the sampling operation in time domain. The sampled signal in frequency domain can also be expressed by finding the Fourier transform of a signal $x_s(t)$.

$$\text{i.e., } x(t) \xrightarrow{\text{FT}} x(f)$$

$$x(f) = \int\limits_{-\infty}^{\infty} x(t) e^{-j2\pi f_s t} dt \qquad (4.4)$$

The frequency domain signal of sampled signal can be obtained in a similar manner.

$$x_s(t) \xrightarrow{\text{FT}} x_s(f)$$

$$x_s(f) = \int\limits_{-\infty}^{\infty} x_s(t) e^{-j2\pi f_s t} dt \qquad (4.5)$$

Substituting Equation (4.3) in Equation (4.5), we get

$$x_s(f) = \int\limits_{-\infty}^{\infty} x(t) g(t) e^{-j2\pi f_s t} dt \qquad (4.6)$$

Substituting the value of $g(t)$ from Equation (4.1) to Equation (4.6)

$$= \int\limits_{-\infty}^{\infty} x(t) \sum_{n=-\infty}^{\infty} C_n e^{j2\pi n f_s t} e^{-j2\pi f t} dt$$

$$= \sum_{n=-\infty}^{\infty} C_n \int\limits_{-\infty}^{\infty} x(t) e^{-j2\pi(f-nf_s)t} dt$$

From the definition of Fourier transform,

$$\int\limits_{-\infty}^{\infty} x(t) e^{-j2\pi(f-nf_s)t} dt = X(f - nf_s)$$

$$x_s(f) = \sum_{n=-\infty}^{\infty} C_n X(f - nf_s) \qquad (4.7)$$

Equation (4.7) can be represented pictorially, as shown in Figure 4.4. It shows sampling in time domain introducing periodicity in the frequency domain, i.e. the same band limited spectrum repeated for every sample interval. This is shown in Figure 4.4(b–d).

Figure 4.4(a) represents the spectrum of band limited modulating signal $x(f)$. Figure 4.4(b) represents the spectrum of sampled signal $x_s(f)$; when $f_s \geq 2f_h$, i.e. sampling frequency is twice the maximum frequency of the

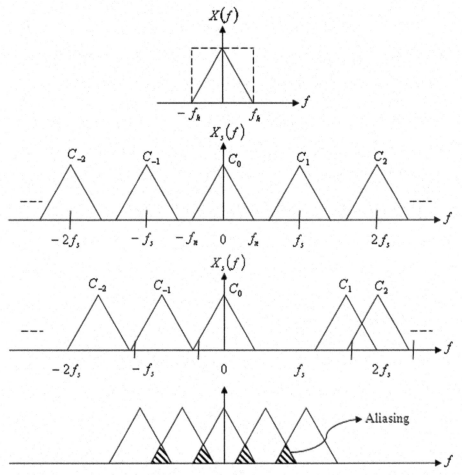

Figure 4.4 (a) Spectrum of sampled signal, (b) Spectrum of sampled signal when $f_s \geq 2f_h$-Over sampling, (c) Spectrum of sampled signal when $f_s = 2f_h$-Critical sampling, (d) Spectrum of sampled signal when $f_s < 2f_h$-Under sampling.

modulating signal. It is also referred to as "Over sampling". The spectrum of $x_s(f)$ is passed to a low pass filter for reconstruction of the signal. A low pass filter allows the spectrum from the band $-f_h$ to $+f_h$ and suppresses all other side bands. This is shown in the figure by a dotted line. Figure 4.4(c) represents the spectrum of a sampled signal when $f_s = 2f_h$. It is also referred to as "Critical sampling". Figure 4.4(d) represents the spectrum of sampled signal; when $f_s < 2f_h$. It is referred to as "under sampling".

The conclusions drawn from the figure are:

- Band limited signal $x(t)$ can be perfectly recovered at the receiver side only when the sampling frequency is $f_s \geq 2f_h$.
- When sampling frequency does not satisfy the above condition, recovery of the original signal at the receiver side is not possible. Also the signals are also affected by adjacent samples. This effect is referred to as "Aliasing effect". Hence, the condition for Aliasing: $f_s \leq 2f_h$.

In other words, for perfect reconstruction of sampling interval, (T_s), is always higher than $1/2T$, i.e. the rate of closer of electronic switch at the transmitter side must satisfy the condition of Nyquist interval.

Sampling Theorem or Nyquist sampling Theorem:

A band limited signal $x(t)$ which has no frequency components above f_h, can be completely specified by samples at a rate greater than or equal to $2f_h$.

$$\text{i.e. } f_s \geq 2f_h \tag{4.8}$$

Perfect recovery of the signal is not possible at the receiver side, when the sampling interval at the transmitter side is not followed.

4.2.2 Detection of PAM Signal

Just passing the spectrum of $x_s(f)$ to low pass filter may be done for reconstructing the signal. Low-pass filter allows the spectrum from the band $-f_h$ to $+f_h$ and suppresses all other side bands. This is shown in the Figure 4.4(a) by dotted line. Figure 4.5 illustrates the detection circuit of PAM signal.

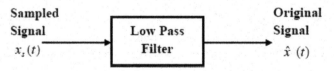

Figure 4.5   Detection of PAM signal.

4.3 Pulse Width Modulation

In PWM, instantaneous pulse width at each pulse duration is proportional to the amplitude of modulating signal.

4.3.1 Generation of PWM Signal

Step 1: Generate PAM signal

PAM signal $x_1(t)$ may be obtained when $x(t)$ and $g(t)$ is applied to PAM generation circuit. This is shown in Figure 4.6.

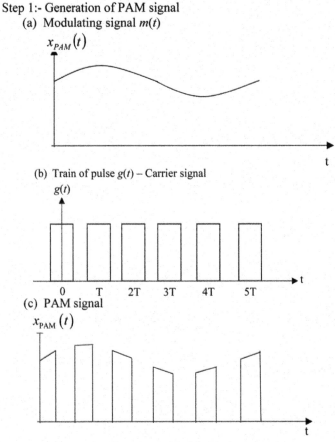

Figure 4.6 Generation of PAM signal.

Step 2: Generate sawtooth waveform

Another set of pulses $p(t)$ with constant amplitude may now be generated, it has a triangular shape This is shown in Figure 4.7.

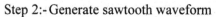

Step 2:- Generate sawtooth waveform

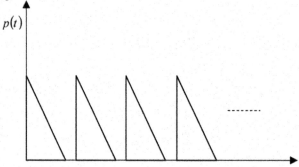

Figure 4.7   Generation of triangular pulse.

Step 3:

Now the waveform $x_{PAM}(t)$ and $p(t)$ are added up. This is shown in Figure 4.8.

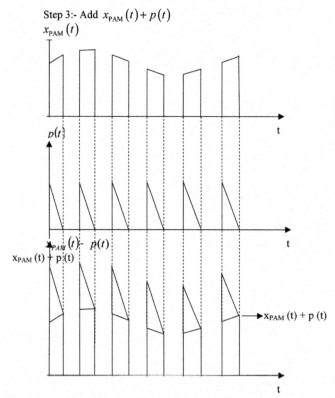

Figure 4.8   Waveform of $x_{PAM}(t) + p(t)$.

Step 4: This waveform (Figure 4.8) may now be sliced at some point, i.e. an arbitrary threshold (V) is to be chosen. This is shown in Figure 4.9.

Step 4:- $x_{PAM}(t) + p(t)$

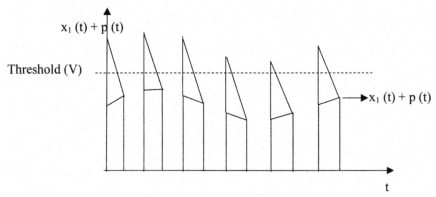

Figure 4.9 Slicing of waveform.

Step 5: A comparator circuit is taken as shown in Figure 4.10. The input for the comparator are $x_{PAM}(t) + p(t)$ and threshold fixed (V) in the previous steps. The output of the comparators are,

$$x_{PAM}(t) + p(t) > \text{Threshold}(V) = 1$$
$$x_{PAM}(t) + p(t) < \text{Threshold}(V) = 0$$

(4.9)

Step 5:-

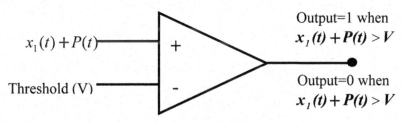

Figure 4.10 Comparator circuit.

Step 6: The comparator output is a PWM waveform which is shown in Figure 4.11.

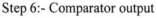

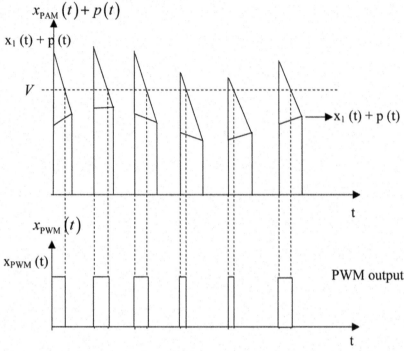

Figure 4.11   PWM waveform.

4.3.2 Detection of PWM

The spectrum of PWM signal is mathematically given as,

$$x_{\text{PWM}}(t) = -\frac{1}{\pi}h\sum\frac{1}{m}J_0\left(m\pi\beta\right)\left\{\left[\sin 2\pi\left(mf_s+mf_m\right)t+\frac{m\pi}{2}\right]\right.$$

$$\left.+\sin\left[\sin 2\pi\left(mf_s-mf_m\right)t+\frac{m\pi}{2}\right]\right\}+\ldots \qquad (4.10)$$

It shows that, similar to FM, PWM spectrum also consists of infinite sidebands around the desired modulating signal. This is shown in Figure 4.12. Hence, with the help of a low pass filter the original signal at receiver side can be detected easily the spectrum of $x_{\text{PWM}}(f)$ may be passed to low pass filter for reconstruction. The low pass filter allows the spectrum from the band $-f_h$ to $+f_h$ and suppresses all other side bands. The detection circuit of PWM is illustrates in Figure 4.13.

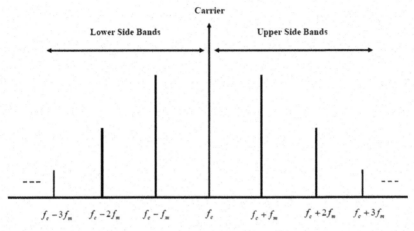

Figure 4.12 Spectrum of PWM signal.

Figure 4.13 Detection of PWM signal.

4.4 Pulse Position Modulation

The position of the pulse in PPM is changed with respect to the position of the reference pulse.

4.4.1 Generation of PPM

PWM signal, it is well known, can be generated with the help of PAM. Similarly PPM signal can be generated by the PWM signal. The steps for generating PPM signal are as follows,

Step 1: Generate PWM signal. It is shown in Figure 4.14.

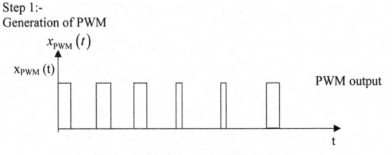

Figure 4.14 Generation of PWM signal.

Step 2: Arrow in the PWM signal indicates the trailing edge of the pulse. This is shown in Figure 4.15 (a).

Step 3: Consider a mono stable multi-vibrator, the input for the mono stable multi-vibrator circuit are PWM signal and triggering pulse. This mono-stable multi-vibrator triggered at the trailing edge of the pulse.

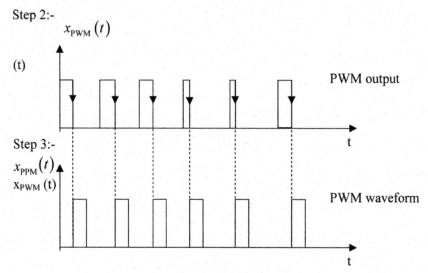

Figure 4.15   (a) Trailing edge of PWM (b) PPM waveform.

Step 4: Operation of mono stable multi-vibrator

Mono stable multivibrators are electronic circuits. When triggered by a triggering pulse, they go to a set or a stable state and come to the original state after a brief period. This is shown in Figure 4.16. The output waveform has a constant width and amplitude but its positions differ or vary according to the triggering pulse.

4.4.2 Detection of PPM

Let us assume the modulating signal as $m(t) = A \cos 2\pi f_m t$. The spectrum of PPM signal is given by the expression,

$$x_{\text{PPM}}(t) = f_s - \beta f_m \sin 2\pi f_m t + \sum \sum \ldots \qquad (4.11)$$

This expression shows the spectrum as consisting of a derivative of modulating signal and higher order terms.

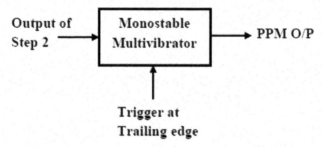

Figure 4.16 Operation of mono stable multivibrator.

Hence the PPM signal can be detected by a low pass filter followed by an integrator circuit. This is shown in Figure 4.17. The spectrum of $x_{PPM}(f)$ has to be passed on to a low pass filter for reconstruction of the signal. Low pass filter allows the spectrum from the band $-f_h$ to $+f_h$ and suppresses all other side bands. These outputs are then given to an integrator, which detects the original signal at the receiver side.

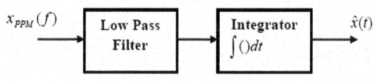

Figure 4.17 Detection of PPM.

Following Table 4.1 lists the comparison of pulse modulation schemes.

Table 4.1 Comparison of pulse modulation schemes

S.No	Parameter	PAM	PWM	PPM
1	Type of carrier	Train of pulses	Train of pulses	Train of pulses
2	Variable characteristic of pulsed carrier	Amplitude	Width	Position
3	Bandwidth requirement	Low	High	High
4	Noise immunity	Low	High	High
5	Information is contained in	Amplitude variation	Width variation	Position variation
6	Transmitted power	Varies with amplitude of pulses	Varies with variation in width of pulse	Constant
7	Synchronization pulse	Not needed	Not needed	Necessary
8	Complexity level of generation and detection	Complex	Easy	Complex

4.5 Pulse Code Modulation

PCM is a type of digital pulse modulation technique, since the output of a PCM system is a digital signal. Continuous time signals are converted into discrete time signals using the sampling process. A digital signal is obtained when the sampled signals are quantized and encoded. Figure 4.18 describes the basic block diagram of the PCM system. It is also called an analog to digital converter. Sampler block accepts a continuous time signal as input and the output of the sampler circuit is discrete in time and continuous in the amplitude format. Quantizer converts a discrete in time and continuous in amplitude signal into discrete in time and amplitude signal. An encoder block assigns the digital word to individual discrete amplitude levels. The individual blocks of PCM are discussed in the sections that follow.

Sampler

The sampler converts the continuous time signal into a discrete time signal. This can also be performed by sampling circuits. Sampling circuits are made of electronic switches. The inputs for the switches are continuous time signals and train of pulses. The switch produces an output only when the switch is closed. Hence the output of the switching circuits are not continuous in nature. Figure 4.19 shows the sampler as a switching circuit. The continuous time signals are sampled at "Nyquist rate", otherwise recovery of the signal at receiver side is not possible. The necessary condition for signal sampling at sampler may be derived. The condition in frequency domain can be obtained easily. Figure 4.20 shows the input waveform, carrier signal and sampled output waveform of the sampling operation.

Let $x(t)$ be the modulating signal (continuous)

$g(t)$ be the carrier (pulse) signal and $x_s(t)$ sampled signal.

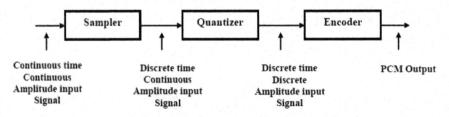

Figure 4.18   Basic blocks of PCM system.

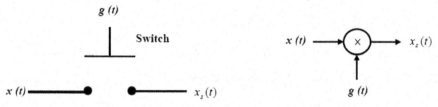

Figure 4.19   Modeling of a sampler as a switch.

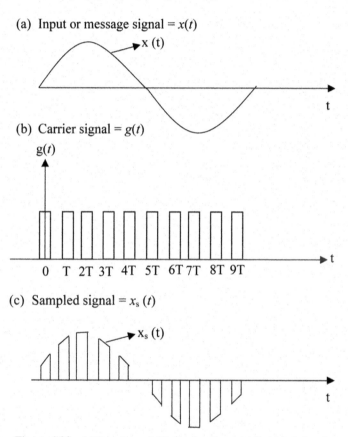

Figure 4.20   (a) Input signal (b) Carrier signal (c) Sampled signal.

Now $g(t)$ are periodic train of pulses. Any periodic signal can be represented by a Fourier series. Where representation of the carrier signal $g(t)$ is given by,

$$g(t) = \sum_{n=-\infty}^{\infty} C_n e^{j2\pi n f_s t} \tag{4.12}$$

where C_n is a nth Fourier coefficients and it can be expressed as, Sampling operation:

$$C_n = \frac{1}{T} \int_{-\frac{f_s}{2}}^{\frac{f_s}{2}} g(t) e^{-j2\pi n f_s t} \tag{4.13}$$

f_s is the sampling frequency.

The output of an electronic switch or a sampler circuit is given as

$$x_s(t) = x(t) g(t) \tag{4.14}$$

This expression describes the sampling operation in a time domain. The sampled signal in frequency domain can also be expressed by finding the Fourier transform of a signal $x_s(t)$.

i.e., $x(t) \xrightarrow{FT} x(f)$

$$x(f) = \int_{-\infty}^{\infty} x(t) e^{-j2\pi f_s t} dt \tag{4.15}$$

The frequency domain signal of the sampled signal can be obtained in a similar manner.

$$x_s(t) \xrightarrow{FT} x_s(f)$$

$$x_s(f) = \int_{-\infty}^{\infty} x_s(t) e^{-j2\pi f_s t} dt \tag{4.16}$$

Sub (4.3) in (4.5), we get

$$x_s(f) = \int_{-\infty}^{\infty} x(t) g(t) e^{-j2\pi f_s t} dt \tag{4.17}$$

Sub value of $g(t)$ from Equation (4.12) to Equation (4.17)

$$= \int_{-\infty}^{\infty} x(t) \sum_{n=-\infty}^{\infty} C_n e^{j2\pi n f_s t} e^{-j2\pi f t} dt$$

$$= \sum_{n=-\infty}^{\infty} C_n \int_{-\infty}^{\infty} x\,(t)\, e^{-j2\pi(f-nf_s)t} \mathrm{d}t$$

From the definition of Fourier transform,

$$\int_{-\infty}^{\infty} x\,(t)\, e^{-j2\pi(f-nf_s)t} \mathrm{d}t = X(f - nf_s)$$

$$x_s\,(f) = \sum_{n=-\infty}^{\infty} C_n X\,(f - nf_s) \qquad (4.18)$$

Pictorial representation of Equation (4.18) can be done, as shown in Figure 4.21. It shows sampling in time domain introducing periodicity in the frequency domain. i.e. the same band limited spectrum repeated for every sample interval. This is shown in Figure 4.21(b–d).

Figure 4.21(a) represents the spectrum of band limited modulating signal $x\,(f)$. Figure 4.21(b) represents the spectrum of sampled signal $x_s\,(f)$; when $f_s \geq 2f_h$, i.e., sampling frequency is twice the maximum frequency of modulating signal. It is also referred to as "Over sampling". The spectrum of $x_s\,(f)$ is passed to a low pass filter for reconstruction of the signal. Low pass filter allows the spectrum from the band $-f_h$ to $+f_h$ and suppress all other side bands. This is shown in the Figure 4.21(a) by a dotted line. Figure 4.21(c) represents the spectrum of a sampled signal when $f_s = 2f_h$. It is also referred to as "Critical sampling". Figure 4.21(d) represents the spectrum of sampled signal; when $f_s \leq 2f_h$. It is referred to as "under sampling".

The conclusion drawn from the figure are:

- Band limited signal $x\,(t)$ can be perfectly recovered at the receiver side only when the sampling frequency is $f_s \geq 2f_h$.
- When sampling frequency does not satisfy the above condition, recovery of the original signal at the receiver side is not possible. Also the signals are affected by adjacent samples. This effect is referred to as "Aliasing effect". Hence the condition for Aliasing: $f_s \leq 2f_h$.

In other words, for perfect reconstruction of sampling interval (T_s), is always higher than $\frac{1}{2T}$, i.e., rate of closure of electronic switch at the transmitter side must satisfy the condition of Nyquist interval.

Sampling Theorem or Nyquist sampling Theorem:

A band limited signal $x\,(t)$ which has no frequency components above f_h, can be completely specified by samples at a rate greater than or equal to $2f_h$.

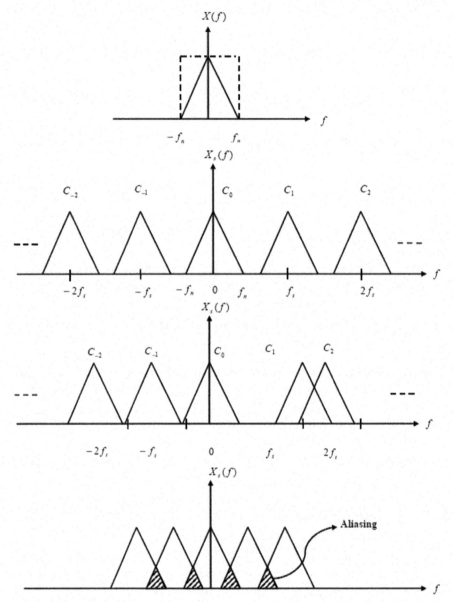

Figure 4.21   (a) Spectrum of sampled signal, (b) Spectrum of sampled signal when $f_s \geq 2f_h$-Over sampling, (c) Spectrum of sampled signal when $f_s = 2f_h$-Critical sampling, and (d) Spectrum of sampled signal when $f_s < 2f_h$-Under sampling.

$$i.e. \ f_s \geq 2f_h \qquad\qquad (4.19)$$

When the sampling interval at transmitter side is not followed, a perfect recovery of the signal at receiver side is not possible.

Quantizer

It converts a discrete-time, continuous-amplitude signal into an discrete time and discrete amplitude signal. Quantization is a process in which the amplitude of each sample is rounded off to the nearest permissible level. The process of quantization introduces a quantization error.

Quantization error

The difference between the original amplitude level to round off amplitude value or the quantized value is called as quantization error.

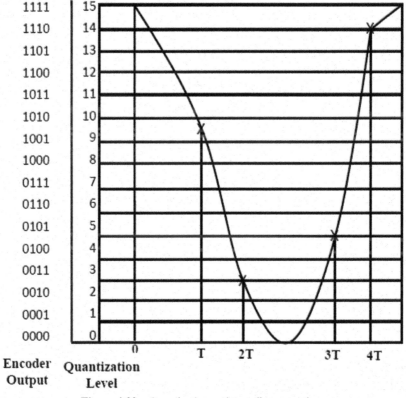

Figure 4.22 Quantization and encoding operation.

For example, in the figure, at time instant T, the original amplitude of signal is 9.8, but this value is rounded off to the nearest level in the quantization process as 10.

Now the quantization error = $10 - 9.8 = 0.2$.

Sampling interval is to be higher for reducing the error in quantization.

Encoder: It represents each permissible level into an equivalent digital word. For example, in the figure, at time instant 2T, output of quantizer is 3V. Now the equivalent digital word for 3V is 0011. This type of conversion is referred to as an encoding operation. This is shown in Figure 4.22.

The number of bits per encoder output is based on the number of quantization level. In Figure 4.22, the total number of quantization levels is 16. Hence, the number of bits to represent the level is four. ($\because 2^4 = 16$).

5

Data Communication

5.1 Introduction

Exchange of digital data between two computers is referred to as data communication. This process requires following of a set of rules that are referred to as "Protocols".

Types of Protocols
Connection oriented and
Connection-less protocol

In the connection oriented protocol, the data received at receiver are followed by an acknowledgement of the signal to the transmitter. Hence, the transmitter knows the condition of data transmission. It is reliable therefore.

In the connection-less protocol, there is no acknowledgement from the receiver to the transmitter, when the receiver receives some information. Hence, it is unreliable.

History of data communication
- The history of data communication began in the pre-industrial age, with information transmitted over short distances using smoke signals, torches, flashing mirrors, signal flags, and semaphore flags.
- In 1838, Samuel Morse and Alfred Veil invented the Morse code telegraph system.
- In 1876, Alexander Graham Bell invented the telephone.
- In 1910, Howard Krum developed the start/stop synchronization.
- In 1930, development of ASCII transmission code began.
- In 1945, Allied governments developed the first large computer.
- In 1950, IBM released its first computer IBM 710.
- In 1960, IBM released the first commercial computer named IBM 360.

- Internet was introduced in the year of 1967 by advanced research project agency (ARPANET) in the USA.
- TCP/IP protocol suite was developed in the year 1975.
- In 70s and 80s, the main thrust in wide area networking (WAN) was to put data on voice circuits using modem and on ISDN lines.
- In 90s, the trend was reversed. Major efforts were on putting voice over data using:
  - Voice over frame relay
  - Voice over Internet
  - Voice over ATM, etc.

Types of data communication
Transmission of data between two computers can be accomplished by either serial or parallel mode of communication.

Parallel transmission
In parallel transmission, multiple bits are sent with each clock period. This is shown in Figure 5.1.

Transmission of eight data needs eight different transmission lines from the transmitter to the receiver. Hence, exchanging a large amount of data needs

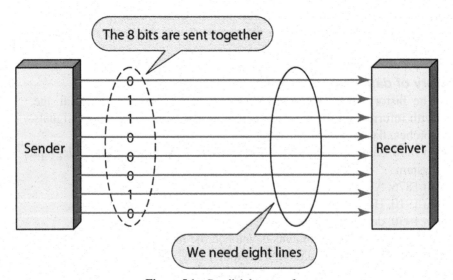

Figure 5.1   Parallel data transfer.

a large number of data lines. Therefore, the parallel mode is suitable only for communication over a short distance.
Example: data transmissi on from one IC to another IC.

Serial mode
In this mode, only one bit is sent for each clock cycle. It was shown in Figure 5.2.

In this Figure 5.2, the sender and the receiver are computers, which can store the data in a parallel mode. Hence, conversion to serial mode has to be done before transmission of the data. Serial mode of communication is efficient when the data are transmitted over longer distances.
Example: Internet.

5.2 Standards and Organization

Standard
An object or procedure considered by an authority or by general consent as a basis of comparison.

Why standards?
It is required to allow for interoperability between equipment.

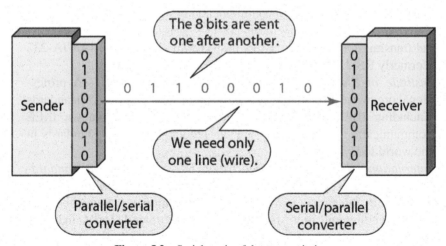

Figure 5.2   Serial mode of data transmission.

Advantages
- Ensures a large market for equipment and software.
- Allows products communication from different vendors.

Disadvantages
- Freeze technology.
- May be multiple standards for the same thing

Key Organizations
Standards organizations provide forums for discussion, helping turn discussion into formal specifications. Most of them have specific processes for turning ideas into formal standards. They all iterate through several rounds of organizing ideas, discussing the ideas, developing draft standards, voting on all or certain aspects of the standards, and finally release the completed standard formally to the public.

Some of the better-known standards organizations follow:

- *International Organization for Standardization (ISO)*—An international standards organization is responsible for a wide range of standards including those relevant to networking. This organization is responsible for the OSI reference model and the OSI protocol suite.
- *American National Standards Institute (ANSI)*—This is the coordinating body for voluntary standards groups within the USA. ANSI is a member of ISO. ANSI's best-known communications standard is FDDI.
- *Electronic Industries Association (EIA)*—A group that specifies electrical transmission standards. EIA's best-known standard is EIA/TIA-232 (formerly RS-232).
- *Institute of Electrical and Electronic Engineers (IEEE)*—A professional organization that defines network standards. IEEE LAN standards (including IEEE 802.3 and IEEE 802.5) are the best-known IEEE communication standards and are the predominant LAN standards in the world today.
- *International Telecommunication Union Telecommunication Standardization Sector (ITU-T)(formerly the Committee for International Telegraph and Telephone [CCITT])*—An international organization that develops communication standards. The best-known ITU-T standard is X.25.
- *Internet Activities Board (IAB)*—A group of internetwork researchers who meet regularly to discuss issues pertaining to the Internet. This board

sets the policies for the Internet through decisions and assignment of task forces to various issues.

- Some *Request for Comments* (RFC) documents are designated by the IAB as Internet standards, including *Transmission Control Protocol/Internet Protocol* (TCP/IP) and the *Simple Network Management Protocol* (SNMP).

5.3 Serial Communication Interface

Interface

Interface is a separate hardware mechanism used for connecting input and output unit (peripheral) with a computer. There are two types of interfacing in data communication. They are serial and parallel communication interface mechanism.

Serial communication interface

The examples for serial communicating devices are keyboard and mouse. The corresponding interface mechanism for serial communication is RS-232 C.

RS-232 C-Serial communication Interface standard:

The main components involved in RS-232C standard are data terminal equipment (DTE) and data circuit-terminating equipment (DTE). Simply stated, DTE is computer and data communication equipment (DCE) is MODEM.

RS-232C standard defines the electrical, the mechanical, and the function specifications of the interface between DTE and DCE.

Mechanical specifications:

RS-232C cable is a 25-pin connector available in the name of DB-25. The length of the cable may not exceed 15 m. Another version of RS-232C cable is DB-9.

Electrical specification:

RS-232C follows the negative logic system, i.e. binary 1 is represented by –3V to –15V and binary 0 is represented by +3V to +15V. It allows a maximum bit rate of 20 Kbps.

Functional specification

Only 4 pins out of 25 pins are used for data transmission, the remaining 21 lines being used for control, timing, grounding, and testing. Functional

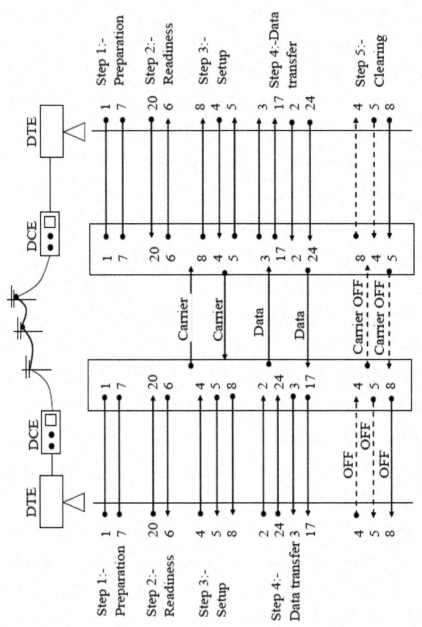

Figure 5.3   Functional specification of RS-232-C serial interface standard.

specifications of RS-232C cable can be given by five different steps. Figure 5.3 shows the synchronous full-duplex transmission mode. Here, DTE represents the computer and DCE refers to MODEM.

Step 1: Preparation of interface for transmission

Two grounding circuits with pin numbers 1 and 7 are active between both the sending computer/modem (left) and receiving computer/modem (right) combination.

Step 2: To ensure that all four devices are ready for transmission

First sending DTE sends the DTE ready information to DCE via pin number 20. To answer that, DCE sends DCE a ready message via pin number 6. The same operation is repeated at the destination side also.

Step 3: Set up a physical connection between sending and receiving modems

This operation can be done as follows: DTE activates pin number 4 and sends request to DCE to send message. The DCE then transmits a carrier signal to the receiving modem. When the receiver modem detects the carrier signal, it activates pin number 8-received signal line detector. It tells its computer that the transmission is about to begin. After sending the carrier signal, the sending DCE activates pin number 5, sending its DTE to clear-t-send message. The destination computer and modem also perform the same step.

Step 4: Data transmission stage

The transmitting computer transfers its data stream to a modem via pin number 2, accompanied by a timing pulse at pin number 24. The transmitter side modem converts the digital data to an analog signal and sends it out over the network.

The receiving modem converts the received analog signal into a digital signal and passes it on to its computer via pin number 3, accompanied by the timing pulse at pin number 17.

Step 5: Clearing

Once the data transmissions are completed, both the computers deactivate their request to send circuits; the modems turn off their carrier signals, received signal line detectors, and clear to send circuits.

5.4 Parallel Communication Interface

Parallel communication interfaces are used for a parallel transfer of the data. Example: Data transfer from PC to printer. The example for parallel communication interface is "Centronics printer interface".

Centronics printer interface:
Pin details:
These are available in the form of female connectors with 25 pins on PC side. The number of pins in the printer side is 36 and the connector is named as centronics. Figure 5.4 shows the system configuration of Centronics interface.

Ground lines:
It consists of 8 ground lines and a twisted pair of cables providing a sufficient shielding of signals. Also at the printer side, every alternate wire is ground. Hence, overall noise effects are minimum in this type of interface.

Data lines:
The total number of data lines is eight and is available in the form of $D_0 - D_7$. All these lines are unidirectional lines (i.e. transfer data from PC to printer only). Data characters are in the form of seven-bit ASCII format.

Control lines:
Signal from PC to printer:
The signals, 12 in number are available from PC to printer communication. Out of these, 8 lines are data lines and the remaining four are control lines.

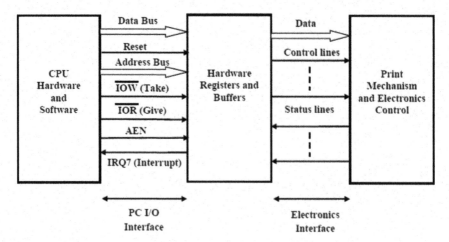

Figure 5.4   Printer controller.

Some of the control lines are $\overline{\text{STROBE}}$, $\overline{\text{INIT}}$, $\overline{\text{SLCTIN}}$, $\overline{\text{AUTOFEEDIN}}$. Figure 5.5 shows the control lines and status lines of the printer controller.

$\overline{\text{STROBE}}$: Printer should take the data, when this signal is low.

$\overline{\text{INIT}}$: When $\overline{\text{INIT}}$ is low, the printer resets the electronic logic and clears the printer buffer.

$\overline{\text{SLCTIN}}$: It is an interface enable signal, when it is low, printer responds to signals from the controller.

$\overline{\text{AUTOFEEDIN}}$: After printing, printer provides one line feed automatically when this signal is low. This type of line feed is known as hardware line feed.

Signals from printer to PC:
Status signal:
Five status signals are available in all the Centronics interface. It is used for informing the status of the printer to the PC. Some of the status signals are: $\overline{\text{ACK}}$, BUSY, PE, SLCT and $\overline{\text{ERROR}}$.

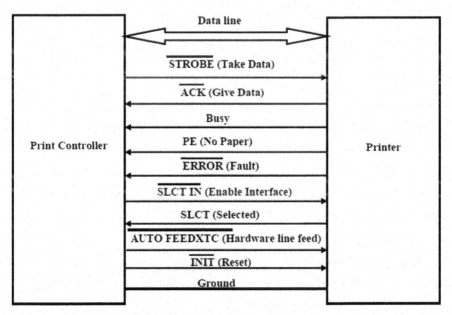

Figure 5.5 Signals in centronics interface.

$\overline{\text{ACK}}$: It is an acknowledgement signal to $\overline{\text{STROBE}}$ signal from the PC. When activated, it indicates printer having received data sent by the PC and readiness of the printer to accept the new data byte.

BUSY : It is an active high signal. When it is high, it indicates that the printer is busy and and cannot receive data.
This signal is high under any of the following situations:

(i) On receiving $\overline{\text{STROBE}}$ active.
(ii) During printing operation.
(iii) Printer in offline state.
(iv) Printer senses some error condition.

PE : It indicates no paper in the printer.
SLCT : It indicates that printer is selected and logically connected to PC.

$\overline{\text{ERROR}}$: It indicates error condition in printer.

Reason for $\overline{\text{ERROR}}$ signal:

(i) Mechanical fault.
(ii) The printer in offline state.
(iii) No paper in the printer.

5.5 Error Detection and Correction Techniques

Error Detection:

Data transfer from one computer to another can be corrupted during transmission. Errors must be detected and corrected to ensure reliable communication. In general, errors are classified into two types. They are: i) Single bit error and ii) Burst error.

Single bit error:

During the data transmission from sender to receiver, only one bit in the data unit has changed. Figure 5.6 shows the single bit error.

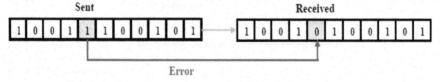

Figure 5.6   Single bit error.

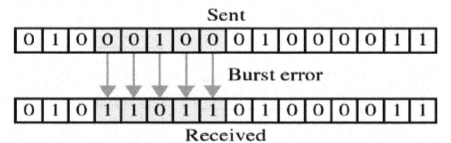

Figure 5.7   Burst error.

Burst error:

When two or more bits in the data unit have changed, it is referred to as a burst error, which most likely happens in serial transmission. Figure 5.7 shows the burst error.

5.5.1 Error Detection Techniques

- Error detection involves decision on the correctness or otherwise of the received data without having a copy of the original message.
- Error detection techniques use the concept of "redundancy", which means addition of extra bits to detect error at the receiver.

Four types of redundancy check can be used at the receiver end.
They are:

1. Vertical redundancy check.
2. Longitudinal redundancy check
3. Cyclic redundancy check and
4. Checksum

1. Vertical redundancy check (VRC):
In this technique, a redundant bit is also called a parity bit and it is appended with the data unit at transmitter side to make the number of ones even. This is shown in Figure 5.8.

Suppose the sender wants to send a character "*a*", whose ASCII code equivalent binary digit is 1100001. Then, even a parity generator at the transmitter side counts the number of ones in the data unit. If the number of ones is odd, "1" is appended as a parity bit to a data unit. Otherwise "0" is to be appended as a parity bit to a data unit. In the given example, the number of one's is odd. Hence, the parity generator generates "1" as a parity bit and appends it to data unit.

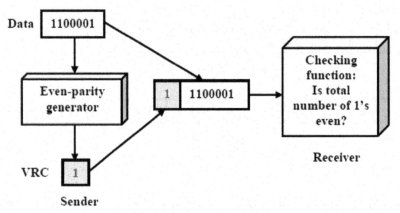

Figure 5.8   Vertical redundancy check.

At the receiver side, the number of one's is counted by the even parity checking function. If the number of one's is even, the receiver accepts the received data; otherwise, it discards such data.

VRC can detect single bit errors as also burst errors as long as the total number of bits changed is odd (1, 3, 5, etc.).

2. Longitudinal redundancy check (LRC):
A block of bits is organized in a table format (row and columns) in LRC. For example, when there is a plan to send a group of 24 bits from the transmitter to the receiver. A data unit is organized in a table format as three rows and eight columns. A parity bit for each column is calculated as a new row of eight bits created, which are parity bits for the whole data block. The first parity bit in the fourth row is calculated based on all the first bits of the first column. The second parity bit is calculated based on all second bits of second column.

Example:

Transmitter side:

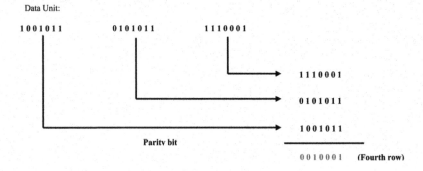

Now the data transmitted by transmitter is:
1 0 0 1 0 1 1   0 1 0 1 0 1 1   1 1 1 0 0 0 1   0 0 1 0 0 0 1

Receiver side:

The received data units are arranged in the table format. Now a parity bit is calculated for each column and a new row of eight bits is created which is now fifth row, when the fifth row has zero as a value in all the columns. Then, the received information is correct. Otherwise, the receiver rejects the entire data unit.

Example:

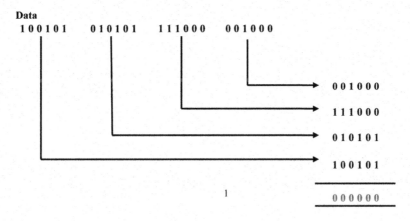

So, there is no error.

LRC increases the likelihood of detecting burst errors. If two bits in a data unit are damaged and such two bits are exactly the same, the positions in another data unit are also damaged, and the LRC checker does not detect an error.

3) Cyclic redundancy check (CRC):

It is based on the binary division technique. Hence a common divisor is used for dividing the data unit both at transmitter and receiver.

The CRC technique involves the following procedures:

Transmitter side:

Assuming the data unit is 11010011101100; and divisor is 1 0 1 1.

n-1 zeros are appended at the end of the data unit, where n is the total number of bits in divisor. In the given problem, the total number of bits in the divisor is four. Therefore, three zeros are append to the data unit.

The division operation is given below:

11010011101100 000	← – –	input right padded by 3 bits
1011	← – –	divisor
01100011101100 000	← – –	result (note the first four bits are the XOR with the divisor beneath, the rest of the bits
1011	← – –	divisor ...
00111011101100 000		
1011		
00010111101100 000		
1011		
00000001101100 000	← – –	note that the divisor moves over to align with the next 1 in the dividend (since quotient for that step as zero)
1011	← – –	(in other words, it doesn't necessarily move one bit per iteration)
00000000110100 000		
1011		
00000000011000 000		
1011		
00000000001110 000		
1011		
00000000000101 000		
101 1		
00000000000000 100	← – –	remainder (3 bits). Division algorithm stops here as quotient is equal to zero.

Now data transmitted by the transmitter is Data unit + remainder.

i.e. 11010011101100 100

Receiver side:

Now the received data unit are divided by the divisor. When the division operation produces zero remainder, the received information is not affected with error; otherwise, the received information is considered corrupted by noise.

11010011101100 100	◄ – –	input with check value
1011	◄ – –	divisor
01100011101100 100	◄ – –	result
1011	◄ – –	divisor ...

01100011101100 100		

```
          1011
01100011101100 100
          101 1
      -----------
                      0  ◄ -- remainder
```

CRC generator is usually represented by polynomials. Example: $x^7 + x^5 + x^2 + x + 1$.

CRC can detect all burst errors that affect an odd number of bits.

4. Checksum:
Transmitter side:

- The data unit is divided into k sections, each with n bits.
- All sections are added together using 1's complement to get the sum.
- The sum is complemented and becomes the *checksum*.
- The checksum is sent with the data.

For example:

Data unit: 1 0 1 1 0 1 0 1   1 1 0 1 0 1 1 1

Divide into two groups of 8 bits each:

```
1 0 1 1 0 1 0 1
1 1 0 1 0 1 1 1
---------------
1 0 0 0 1 1 0 0 (Add by 1's complement method)
```

Now the result is complemented.

The complement:

0 1 1 1 0 0 1 1. Now the result is checksum. It is added with the data unit and sent to receiver.

Receiver side:

- The received data unit is divided into k sections, each with n bits.
- All sections are added together using 1's complement to get the sum.
- The sum is complemented.
- When the result is zero, the data are accepted. Otherwise, they are rejected.

```
1 0 1 1 0 1 0 1
1 1 0 1 0 1 1 1
0 1 1 1 0 0 1 1
---------------
1 1 1 1 1 1 1 1
```

Complement 0 0 0 0 0 0 0 0 $\Longrightarrow$ no error

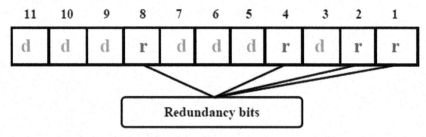

Figure 5.9   Position of redundant bit.

5.5.2 Error Correction Technique

Single bit error correction:

When the receiver knows the location of error, it can correct a single-bit error. Hamming code is the best example of a single-bit error correction technique. Let us assume that the number of bits in the data unit is *m*.

The redundant bits are *r*.

Then total transmitted bits is *m* + *r* .

The value of *r* must satisfy

$$2^r \geq m + r + 1.$$

For example, let us assume ASCII format of data unit (i.e. data unit length is 7 bits). The number of redundant bits is 4. The position of redundant bit is shown in Figure 5.9.

Purpose of redundant bits r1, r2, r4, and r8 are shown in Figure 5.10.

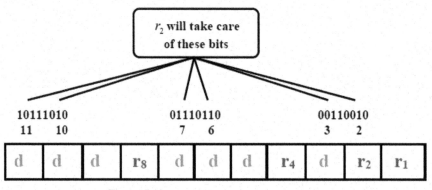

Figure 5.10   (a) Purpose of redundant bit r2.

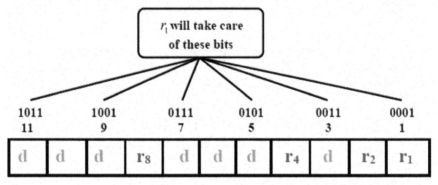

Figure 5.10   (b) Purpose of redundant bit r1.

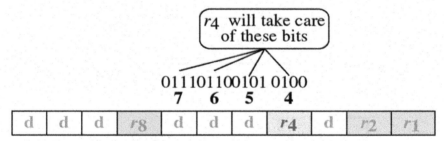

Figure 5.10   (c) Purpose of redundant bit r4.

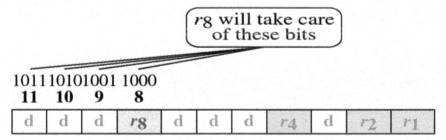

Figure 5.10   (d) Purpose of redundant bits r8.

For example: data to be transmitted by transmitter is: 1 0 0 1 1 0 1.

Then the redundant bits r1, r2, r3, and r4 are calculated using the following procedure:

r1 = bit3 $\oplus$ bit5 $\oplus$ bit7 $\oplus$ bit9 $\oplus$ bit11

r2 = bit3 $\oplus$ bit6 $\oplus$ bit7 $\oplus$ bit10 $\oplus$ bit11

r4 = bit5 $\oplus$ bit6 $\oplus$ bit7

r8 = bit9 $\oplus$ bit10 $\oplus$ bit11

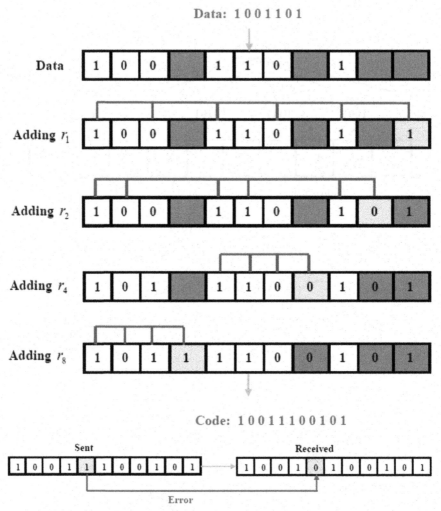

Figure 5.11   Bit affected by noise.

Therefore, data transmitted by transmitter is 1 0 0 1 1 1 0 0 1 0 1.
Receiver side:
Let us assume that the 7th bit is affected by noise. It is shown in Figure 5.11.
Then receiver finds out redundant bits r1, r2, r4, and r8 at the receiver side.
It is shown in Figure 5.12.

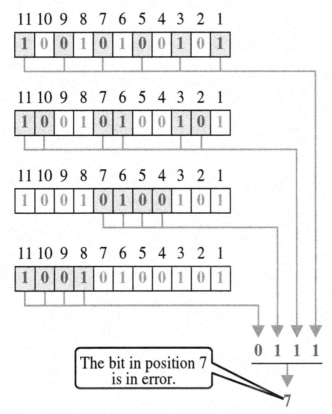

Figure 5.12 Checksum method of error detection.

It shows that, bit position 7 is affected by noise. Hence, the receiver takes the complement of data in the 7th bit position. In this way, we can correct the single bit error.

PART IV

Source Coding and Channel Coding Techniques

6

Source Coding

6.1 Introduction

The basic difference between source coding and channel coding can be explained as:

Source coding: It refers to removal of redundant data or data that are not that important for conveying.

Channel coding: It refers to addition of extra bits in the form of parity bits to user data to help the user to get protection of data from noises. Chapters 6 and 7 cover the various methods of source coding and channel coding techniques used in digital communication.

6.1.1 Discrete Memoryless Source

The basic unit of digital communication system is *"Information source"* or simply *source*. Consider a source S, emitting different symbols $s_1, s_2, s_3 \ldots s_k$ for various time instants, i.e. at time period 1, the source S emits symbol s_1, at time period 2 it emits symbol s_2, and so on. These symbols are statistically independent of each other. They are discrete in nature. Let p_1 be the probability of occurrence of symbol s_1, p_2 the probability of occurrence of symbol s_2, and then p_k the probability of occurrence of symbol s_k. These set of probabilities should satisfy the following condition:

$$\sum_{k=0}^{K-1} p_k = 1.$$

When a source satisfies the above conditions, it is referred as a *discrete memoryless source.*

Concept of Discrete memoryless source

The term discrete refers to source emitting symbols $s_1, s_2, s_3 \ldots s_k$ for various time instants $t_1, t_2, t_3 \ldots t_k$ and these symbols are statistically independent. The term memoryless refers to the output of za source at any time instant depending on the present input and not with any past inputs/outputs.

Information

The amount of information gained by the user after observing symbol s_1 is given by

$$I(s_1) = \log\left(\frac{1}{p_1}\right) \tag{6.1}$$

where p_1 refers the probability of occurrence of symbol s_1.

Similarly, the amount of information gained by the user from the symbol s_k is

$$I(s_k) = \log\left(\frac{1}{p_k}\right) \tag{6.2}$$

where p_k refers probability of occurrence of symbol s_k.

When the base of the logarithm is 2, the unit of information is a *bit*. It can be written as

$$I(s_k) = \log_2\left(\frac{1}{p_k}\right) \text{ bits } \quad k = 0, 1, \ldots K - 1 \tag{6.3}$$

6.2 Entropy

The mean value of $I(s_k)$ over the source alphabet S is given by

$$H(S) = E\left[I\left(s_k\right)\right]$$

$$= \sum_{k=0}^{K-1} p_k I\left(s_k\right)$$

$$= \sum_{k=0}^{K-1} p_k \log_2\left(\frac{1}{p_k}\right) \tag{6.4}$$

The quantity $H(s)$ is called the *entropy* of a discrete memoryless source with source alphabet S.

Definition:

Entropy is a measure of the average information per source symbol. It depends only on the probabilities of the symbols.

Some results:

$H(S) = 0$, if and only if the probability $p_k = 1$.

Extension of discrete memoryless source

$$H(S^n) = nH(S) \tag{6.5}$$

6.2.1 Source-Coding Techniques

The various source-coding techniques are:

1. Shannon Fano coding
2. Huffman coding
3. Lempel Ziv coding
4. Prefix coding

6.3 Shannon Fano Coding

In Shannon Fano coding, a small number of bits are assigned to higher probable events and a large number of bits are assigned to smaller probable events.

Steps involved in Shannon Fano Coding

1. The source symbols are written in decreasing probabilities.
2. The message set is partitioned into two most equi-probable groups [A1] and [A2]. The probability value of group [A1] should be approximately equal to the value of group [A2], i.e. [A1]>=[A2] or [A1]<=[A2].
3. After partitioning, "0" is assigned to each message contained in group [A1] and "1" to each message contained in group [A2].
4. The same procedure is repeated for the groups [A1] and [A2], i.e. Group [A1] will be divided into two equi-probable groups as [A11] and [A12] and group [A2] will be divided into two equi-probable groups as [A21] and [A22]. The codewords in [A11] start with 00, [A12] starts with 01; [A21] start with 10 and [A22] start with 11.
5. The same procedure is repeated until each group contains only one message.
6. The coding efficiency for Shannon Fano coding is given by the expression
   $\eta = H(S)/\overline{L}$, where $H(S)$ is the entropy of the source $S. H(S) = \sum_{k=0}^{K-1} p_k \log_2(1/p_k)$ and $\overline{L}$ is the average codeword length.

$\overline{L} = \sum_{k=0}^{K-1} p_k l_k$, where l_k is the length of binary codeword assigned to symbol.

7. Redundancy can be calculated by the following formula.
   Redundancy $= 1 - \eta$.

6.4 Huffman Coding

It is a type of source-coding technique in which the average codeword length $(\overline{L})$ approaches the fundamental limit set by the entropy of a discrete memoryless source. Hence, we can obtain maximum efficiency (η) from Huffman code. Therefore, it is called as "*Optimum code*".

Steps Involved in Huffman Coding

1. Arrange the source symbols or messages in the decreasing probability order.
2. The two messages of the lowest probability are assigned 0 and 1. This process is referred to as "*Splitting stage*".
3. These two messages are combined into one new message with probability equal to the sum of two original probabilities. The probability of the new message is placed in the list in accordance with its value. This process is referred to as "*Reduction stage*", since the number of messages or symbols in stage 1 will be reduced by one and this reduction process continues in each stage.
4. The procedure is repeated until we are left with a final list of source symbols, for which a 0 and 1 are assigned.
5. The codeword calculation or encoding process starts from the last reduction stage (rightmost part) to the first reduction stage (leftmost part).

6.5 Solved Problems

6.5.1 Solved Problems on Shannon Fano-Coding Algorithm

1. Find out the Shannon Fano code for a discrete memoryless source with probability statistics

$$\{0.1,\ 0.1,\ 0.2,\ 0.2,\ 0.4\}.$$

Solution:

Step 1:

Arrange the probabilities in the descending order
0.4, 0.2, 0.2, 0.1, 0.1

Step 2:

(i) Divide the total symbol or message into two equi-probable groups as A_1 and A_2. It can be

$$A_1 > A_2 \ or \ A_1 = A_2 \ or \ A_1 < A_2$$

$$\text{Group} A_1 \begin{cases} 0.4 \\ 0.2 \end{cases}$$
$$\text{Group} A_2 \begin{cases} 0.2 \\ 0.1 \\ 0.1 \end{cases}$$

(ii) Assign "0" to group A_1 symbols and assign "1" to group A_2 symbols.

$$\text{Group} A_1 \begin{cases} 0.4 \ \ 0 \\ 0.2 \ \ 0 \end{cases}$$
$$\text{Group} A_2 \begin{cases} 0.2 \ \ 1 \\ 0.1 \ \ 1 \\ 0.1 \ \ 1 \end{cases}$$

(iii) Divide the group A_1 into two equi-probable groups as A_{11} and A_{12}

$$\text{Group} A_1 \begin{cases} 0.4 \ \ 0 \ \text{Group} A_{11} \\ 0.2 \ \ 0 \ \text{Group} A_{12} \end{cases}$$
$$\text{Group} A_2 \begin{cases} 0.2 \ \ 1 \\ 0.1 \ \ 1 \\ 0.1 \ \ 1 \end{cases}$$

(iv) Assign "0" to symbols in group A_{11} and assign "1" to symbols in group A_{12}

$$\text{Group} A_1 \begin{cases} 0.4 \ \ 0 \ \ 0 \ \text{Group} A_{11} \\ 0.2 \ \ 0 \ \ 1 \ \text{Group} A_{12} \end{cases}$$
$$\text{Group} A_2 \begin{cases} 0.2 \ \ 1 \\ 0.1 \ \ 1 \\ 0.1 \ \ 1 \end{cases}$$

(v) Further division of group A_{11} and A_{12} is not possible. Hence, divide the group A_2 into A_{21} and A_{22} by two equi-probable groups.

$$\text{Group}A_1 \begin{cases} \underline{0.4 \quad 0 \quad 0 \quad \text{Group}A_{11}} \\ 0.2 \quad 0 \quad 1 \quad \text{Group}A_{12} \end{cases}$$
$$\text{Group}A_2 \begin{cases} 0.2 \quad 1 \qquad \text{Group}A_{21} \\ 0.1 \quad 1 \\ 0.1 \quad 1 \end{cases} \Bigg\} \text{Group}A_{22}$$

(vi) Assign "0" to symbols in group A_{21} and assign "1" to symbols in group A_{22}

$$\text{Group}A_1 \begin{cases} \underline{0.4 \quad 0 \qquad 0 \, \text{Group}A_{11}} \\ 0.2 \quad 0 \qquad 1 \, \text{Group}A_{12} \end{cases}$$
$$\text{Group}A_2 \begin{cases} 0.2 \quad 1 \quad 0 \quad \text{Group}A_{21} \\ 0.1 \quad 1 \quad 1 \\ 0.1 \quad 1 \quad 1 \end{cases} \Bigg\} \text{Group}A_{22}$$

(vii) Further division of group A_{21} is not possible. Hence, divide group A_{22} into two equi-probable groups A_{221} and A_{222}

$$\text{Group}A_1 \begin{cases} \underline{0.4 \quad 0 \quad 0 \quad \text{Group}A_{11}} \\ 0.2 \quad 0 \quad 1 \quad \text{Group}A_{12} \end{cases}$$
$$\text{Group}A_2 \begin{cases} 0.2 \quad 1 \quad 0 \qquad \text{Group}A_{21} \\ 0.1 \quad 1 \quad 1 \\ 0.1 \quad 1 \quad 1 \end{cases} \Bigg\} \begin{matrix} \text{Group}A_{22}\text{Group}A_{221} \\ \hline \text{Group}A_{222} \end{matrix}$$

(viii) Assign "0" to symbols in group A_{221} and assign "1" to the symbols in group A_{222}.

$$\text{Group}A_1 \begin{cases} \underline{0.4 \quad 0 \quad 0 \quad \text{Group}A_{11}} \\ 0.2 \quad 0 \quad 1 \quad \text{Group}A_{12} \end{cases}$$
$$\text{Group}A_2 \begin{cases} 0.2 \quad 1 \quad 0 \qquad \text{Group}A_{21} \\ 0.1 \quad 1 \quad 1 \quad 0 \\ 0.1 \quad 1 \quad 1 \quad 1 \end{cases} \Bigg\} \begin{matrix} \text{Group}A_{22}\text{Group}A_{221} \\ \hline \text{Group}A_{222} \end{matrix}$$

(ix) Further division of group A_{221} and A_{222} is not possible. Hence, stop the iteration.

Step 3:

Symbol	Probability	Codeword	Length (l)
S_0	0.4	0 0	2
S_1	0.2	0 1	2
S_2	0.2	1 0	2
S_3	0.1	1 1 0	3
S_4	0.1	1 1 1	3

Step 4:

To find out efficiency (η), we must calculate the average codeword length ($\overline{L}$) and entropy $H(S)$.

$$\eta = \frac{H(S)}{\overline{L}}$$

where $\overline{L} = \sum_{k=0}^{4} p_k l_k$

$$= p_0 l_0 + p_1 l_1 + p_2 l_2 + p_3 l_3 + p_4 l_4$$

$$= (0.4)(2) + (0.2)(2) + (0.2)(2) + (0.1)(3) + (0.1)(3)$$

$$= 0.8 + 0.4 + 0.4 + 0.3 + 0.3$$

$$\overline{L} = 2.2 \text{ bits/symbol}$$

$$H(S) = \sum_{k=0}^{4} p_k \log_2 \left(\frac{1}{p_k}\right)$$

$$= p_0 \log_2 \left(\frac{1}{p_0}\right) + p_1 \log_2 \left(\frac{1}{p_1}\right) + p_2 \log_2 \left(\frac{1}{p_2}\right) + p_3 \log_2 \left(\frac{1}{p_3}\right)$$

$$+ p_4 \log_2 \left(\frac{1}{p_4}\right)$$

$$= 0.4 \log_2 \left(\frac{1}{0.4}\right) + 0.2 \log_2 \left(\frac{1}{0.2}\right) + 0.2 \log_2 \left(\frac{1}{0.2}\right) + 0.1 \log_2 \left(\frac{1}{0.1}\right)$$

$$+ 0.1 \log_2 \left(\frac{1}{0.1}\right)$$

$$= 0.4\frac{\log_{10}\left(\frac{1}{0.4}\right)}{\log_{10} 2} + 0.2\frac{\log_{10}\left(\frac{1}{0.2}\right)}{\log_{10} 2} + 0.2\frac{\log_{10}\left(\frac{1}{0.2}\right)}{\log_{10} 2} + 0.1\frac{\log_{10}\left(\frac{1}{0.1}\right)}{\log_{10} 2}$$

$$+ 0.1\frac{\log_{10}\left(\frac{1}{0.1}\right)}{\log_{10} 2}$$

$$= (0.4 \times 1.3219) + (0.2 \times 2.3219) + (0.2 \times 2.3219) + (0.1 \times 3.3219)$$

$$+ (0.1 \times 3.3219)$$

$$= 0.5287 + 0.4643 + 0.4643 + 0.33219 + 0.33219$$

$$= 2.12 \text{ bits/symbol}$$

$$H\left(S\right) = 2.12 \text{ bits/symbol}$$

$$\eta = \frac{H\left(S\right)}{\overline{L}}$$

$$= \frac{2.12}{2.2} = 0.96$$

$$\eta = 96\%$$

2. A discrete memoryless source has five messages S_1, S_2, S_3, S_4 and S_5 with probabilities

$$p\left(S_1\right) = 0.4, p\left(S_2\right) = 0.19, p\left(S_3\right) = 0.16, p\left(S_4\right) = 0.15, p\left(S_5\right) = 0.1.$$

Construct the Shannon-Fano code and calculate the code efficiency.
Solution:
Step 1:

   Arrange the probabilities in the decreasing order
   0.4, 0.19, 0.16, 0.15, 0.1
Step 2:

   (i) Divide the total symbol or message into two equi-probable groups as A_1
   and A_2. It can be $A_1 > A_2$ or $A_1 = A_2$ or $A_1 < A_2$

$$\text{Group} A_1 \begin{cases} 0.4 \\ 0.19 \end{cases}$$

$$\text{Group} A_2 \begin{cases} 0.16 \\ 0.15 \\ 0.1 \end{cases}$$

(ii) Assign "0" to group A_1 symbols and assign "1" to group A_2 symbols.

$$
\text{Group}A_1 \begin{cases} 0.4 & 0 \\ 0.19 & 0 \end{cases}
$$

$$
\text{Group}A_2 \begin{cases} 0.16 & 1 \\ 0.15 & 1 \\ 0.1 & 1 \end{cases}
$$

(iii) Divide the group A_1 into two equi-probable groups as A_{11} and A_{12}

$$
\text{Group}A_1 \begin{cases} 0.4 & 0 \ \text{Group}A_{11} \\ 0.19 & 0 \ \text{Group}A_{12} \end{cases}
$$

$$
\text{Group}A_2 \begin{cases} 0.16 & 1 \\ 0.15 & 1 \\ 0.1 & 1 \end{cases}
$$

(iv) Assign "0" to symbols in group A_{11} and assign "1" to symbols in group A_{12}

$$
\text{Group}A_1 \begin{cases} 0.4 & 0 & 0 \ \text{Group}A_{11} \\ 0.19 & 0 & 1 \ \text{Group}A_{12} \end{cases}
$$

$$
\text{Group}A_2 \begin{cases} 0.16 & 1 \\ 0.15 & 1 \\ 0.1 & 1 \end{cases}
$$

(v) Further division of group A_{11} and A_{12} is not possible. Hence, divide the group A_2 into A_{21} and A_{22} by two equi-probable groups.

$$
\text{Group}A_1 \begin{cases} 0.4 & 0 & 0 \ \text{Group}A_{11} \\ 0.19 & 0 & 1 \ \text{Group}A_{12} \end{cases}
$$

$$
\text{Group}A_2 \begin{cases} 0.16 & 1 & \quad \text{Group}A_{21} \\ 0.15 & 1 & \\ 0.1 & 1 & \end{cases} \Big\} \text{Group}A_{22}
$$

(vi) Assign "0" to symbols in group A_{21} and assign "1" to symbols in group A_{22}

$$
\text{Group}A_1 \begin{cases} 0.4 & 0 & 0 \ \text{Group}A_{11} \\ 0.19 & 0 & 1 \ \text{Group}A_{12} \end{cases}
$$

$$
\text{Group}A_2 \begin{cases} 0.16 & 1 & 0 \quad \text{Group}A_{21} \\ 0.15 & 1 & 1 \\ 0.1 & 1 & 1 \end{cases} \Big\} \text{Group}A_{22}
$$

(vii) Further division of group A_{21} is not possible. Hence, divide the group A_{22} into two equi-probable groups A_{221} and A_{222}.

$$
\text{Group}A_1 \begin{cases} 0.4 & 0 & 0 \;\text{Group}A_{11} \\ 0.19 & 0 & 1 \;\text{Group}A_{12} \end{cases}
$$

$$
\text{Group}A_2 \begin{cases} 0.16 & 1 & 0 & \text{Group}A_{21} \\ 0.15 & 1 & 1 & \quad\Big\} \quad \text{Group}A_{22}\text{Group}A_{221} \\ 0.1 & 1 & 1 & \qquad\qquad \text{Group}A_{222} \end{cases}
$$

(viii) Assign "0" to symbols in group A_{221} and assign "1" to symbols in group A_{222}.

$$
\text{Group}A_1 \begin{cases} 0.4 & 0 & 0 \;\text{Group}A_{11} \\ 0.19 & 0 & 1 \;\text{Group}A_{12} \end{cases}
$$

$$
\text{Group}A_2 \begin{cases} 0.16 & 1 & 0 & & \text{Group}A_{21} \\ 0.15 & 1 & 1 & 0 \;\Big\} & \text{Group}A_{22}\text{Group}A_{221} \\ 0.1 & 1 & 1 & 1 \;\int & \text{Group}A_{222} \end{cases}
$$

(xi) Further division of group A_{221} and A_{222} is not possible. Hence, stop the iteration.

Step 3:

Symbol	Probability	Codeword	Length (l)
S_1	0.4	0 0	2
S_2	0.19	0 1	2
S_3	0.16	1 0	2
S_4	0.15	1 1 0	3
S_5	0.1	1 1 1	3

Step 4: To find out efficiency (η), we must calculate the average codeword length $(\overline{L})$ and entropy $H(S)$.

$$
\eta = \frac{H(S)}{\overline{L}}
$$

where $\overline{L} = \sum_{k=0}^{4} p_k l_k$

$$
= p_0 l_0 + p_1 l_1 + p_2 l_2 + p_3 l_3 + p_4 l_4
$$

$$
= (0.4)(2) + (0.19)(2) + (0.16)(2) + (0.15)(3) + (0.1)(3)
$$

$$= 0.8 + 0.38 + 0.32 + 0.45 + 0.3$$

$$\overline{L} = 2.25 \text{ bits/symbol}$$

$$H(S) = \sum_{k=1}^{5} p_k \log_2 \left(\frac{1}{p_k} \right)$$

$$= p_1 \log_2 \left(\frac{1}{p_1} \right) + p_2 \log_2 \left(\frac{1}{p_2} \right) + p_3 \log_2 \left(\frac{1}{p_3} \right) + p_5 \log_2 \left(\frac{1}{p_5} \right)$$

$$= 0.4 \log_2 \left(\frac{1}{0.4} \right) + 0.19 \log_2 \left(\frac{1}{0.2} \right) + 0.16 \log_2 \left(\frac{1}{0.2} \right)$$

$$+ 0.15 \log_2 \left(\frac{1}{0.1} \right) + 0.1 \log_2 \left(\frac{1}{0.1} \right)$$

$$= 0.4 \frac{\log_{10} \left(\frac{1}{0.4} \right)}{\log_{10} 2} + 0.19 \frac{\log_{10} \left(\frac{1}{0.19} \right)}{\log_{10} 2} + 0.16 \frac{\log_{10} \left(\frac{1}{0.16} \right)}{\log_{10} 2} + 0.15 \frac{\log_{10} \left(\frac{1}{0.15} \right)}{\log_{10} 2}$$

$$+ 0.1 \frac{\log_{10} \left(\frac{1}{0.1} \right)}{\log_{10} 2}$$

$$= (0.4 \times 1.3219) + (0.19 \times 2.3959) + (0.16 \times 2.6438) + (0.15 \times 2.7369)$$

$$+ (0.1 \times 3.3219)$$

$$= 0.5287 + 0.4552 + 0.4230 + 0.4105 + 0.3321$$

$$= 2.14 \text{ bits/symbol}$$

$$H(S) = 2.14 \text{ bits/symbol}$$

$$\eta = \frac{H(S)}{\overline{L}}$$

$$= \frac{2.14}{2.25} = 0.95$$

$$\eta = 95\%$$

3) A discrete memoryless source emits six symbols S_0, S_1, S_2, S_3, S_4 and S_5 with probabilities 0.3, 0.25, 0.05, 0.12, 0.08, and 0.2, respectively. Construct the Shannon Fano code and compute its efficiency.

Solution:

Step 1:

Arrange the probabilities in the descending order
0.3, 0.25, 0.2, 0.12, 0.08, 0.05

Step 2:

Construction of Shannon Fano code:

0.3	0	0		
0.25	0	1		
0.2	1	0		
0.12	1	1	0	
0.08	1	1	1	0
0.05	1	1	1	1

Step 3:

Symbol	Probability	Codeword	Length (l)
S_0	0.3	0 0	2
S_1	0.25	0 1	2
S_2	0.2	1 0	2
S_3	0.12	1 1 0	3
S_4	0.08	1 1 1 0	4
S_5	0.05	1 1 1 1	4

Step 4:

To find out efficiency (η), we must calculate the average codeword length $(\overline{L})$ and entropy $H(S)$.

$$\eta = \frac{H(S)}{\overline{L}}$$

where $\overline{L} = \sum_{k=0}^{5} p_k l_k$

$$= p_0 l_0 + p_1 l_1 + p_2 l_2 + p_3 l_3 + p_4 l_4 + p_5 l_5$$

$$= (0.3)(2) + (0.25)(2) + (0.2)(2) + (0.12)(3) + (0.08)(4) + (0.05)(4)$$

$$= 0.6 + 0.5 + 0.4 + 0.36 + 0.32 + 0.2$$

$$\overline{L} = 2.38 \text{ bits/symbol}$$

$$H\left(S\right) = \sum_{k=0}^{5} p_k \log_2\left(1/p_k\right)$$

$$= p_0 \log_2\left(\frac{1}{p_0}\right) + p_1 \log_2\left(\frac{1}{p_1}\right) + p_2 \log_2\left(\frac{1}{p_2}\right) + p_3 \log_2\left(\frac{1}{p_3}\right)$$

$$+ p_4 \log_2\left(\frac{1}{p_4}\right) + p_5 \log_2\left(\frac{1}{p_5}\right)$$

$$= 0.3 \log_2\left(\frac{1}{0.3}\right) + 0.25 \log_2\left(\frac{1}{0.25}\right) + 0.2 \log_2\left(\frac{1}{0.2}\right)$$

$$+ 0.12 \log_2\left(\frac{1}{0.12}\right) + 0.08 \log_2\left(\frac{1}{0.08}\right) + 0.05 \log_2\left(\frac{1}{0.05}\right)$$

$$= 0.3 \frac{\log_{10}\left(1/0.3\right)}{\log_{10} 2} + 0.25 \frac{\log_{10}\left(1/0.25\right)}{\log_{10} 2} + 0.2 \frac{\log_{10}\left(1/0.2\right)}{\log_{10} 2}$$

$$+ 0.12 \frac{\log_{10}\left(1/0.12\right)}{\log_{10} 2} + 0.08 \frac{\log_{10}\left(1/0.08\right)}{\log_{10} 2} + 0.05 \frac{\log_{10}\left(1/0.05\right)}{\log_{10} 2}$$

$$= (0.3 \times 1.7369) + (0.25 \times 2) + (0.2 \times 2.3219) + (0.12 \times 3.0588)$$

$$+ (0.08 \times 3.6438) + (0.05 \times 4.3219)$$

$$= 0.5210 + 0.5 + 0.4643 + 0.3670 + 0.2915 + 0.2160$$

$$= 2.35 \text{ bits/symbol}$$

$$H\left(S\right) = 2.35 \text{ bits/symbol}$$

$$\eta = \frac{H\left(S\right)}{L}$$

$$= \frac{2.35}{2.38} = 0.9873$$

$$\eta = 98.73\%$$

4. A discrete memoryless source emits five symbols with probabilities 0.15, 0.10, 0.05, 0.15, and 0.55, respectively. Compute efficiency by Shannon Fano algorithm.

Solution:

Step 1:

Arrange the probabilities in the descending order

0.55, 0.15, 0.15, 0.10, 0.05

Step 2:

Construction of Shannon Fano code:

$$
\begin{array}{|llll|}
\hline
0.55 & 0 & & \\
\hline
0.15 & 1 & 0 & 0 \\
\hline
0.15 & 1 & 0 & 1 \\
\hline
0.10 & 1 & 1 & 0 \\
\hline
0.05 & 1 & 1 & 1 \\
\hline
\end{array}
\qquad (6.6)
$$

Step 3:

Symbol	Probability	Codeword	Length (l)
S_0	0.55	0	1
S_1	0.15	1 0 0	3
S_2	0.15	1 0 1	3
S_3	0.10	1 1 0	3
S_4	0.05	1 1 1	3

Step 4:

To find out efficiency (η), we must calculate the average codeword length $(\overline{L})$ and entropy $H(S)$.

$$
\eta = \frac{H(S)}{\overline{L}}
$$

where $\overline{L} = \sum_{k=0}^{4} p_k l_k$

$$
= p_0 l_0 + p_1 l_1 + p_2 l_2 + p_3 l_3 + p_4 l_4
$$

$$
= (0.55)(1) + (0.15)(3) + (0.15)(3) + (0.1)(3) + (0.05)(3)
$$

$$= 0.55 + 0.45 + 0.45 + 0.3 + 0.15$$

$$\overline{L} = 1.9 \text{ bits/symbol}$$

$$H(S) = \sum_{k=0}^{4} p_k \log_2 \left(\frac{1}{p_k}\right)$$

$$= p_0 \log_2 \left(\frac{1}{p_0}\right) + p_1 \log_2 \left(\frac{1}{p_1}\right) + p_2 \log_2 \left(\frac{1}{p_2}\right) + p_3 \log_2 \left(\frac{1}{p_3}\right)$$

$$+ p_4 \log_2 \left(\frac{1}{p_4}\right)$$

$$= 0.55 \log_2 \left(\frac{1}{0.55}\right) + 0.15 \log_2 \left(\frac{1}{0.15}\right) + 0.15 \log_2 \left(\frac{1}{0.15}\right)$$

$$+ 0.1 \log_2 \left(\frac{1}{0.1}\right) + 0.05 \log_2 \left(\frac{1}{0.05}\right)$$

$$= 0.55 \frac{\log_{10}\left(\frac{1}{0.55}\right)}{\log_{10} 2} + 0.15 \frac{\log_{10}\left(\frac{1}{0.15}\right)}{\log_{10} 2} + 0.15 \frac{\log_{10}\left(\frac{1}{0.15}\right)}{\log_{10} 2}$$

$$+ 0.1 \frac{\log_{10}\left(\frac{1}{0.1}\right)}{\log_{10} 2} + 0.05 \frac{\log_{10}\left(\frac{1}{0.05}\right)}{\log_{10} 2}$$

$$= (0.55 \times 0.8624) + (0.15 \times 2.7369) + (0.15 \times 2.7369)$$

$$+ (0.10 \times 3.3219) + (0.05 \times 4.3219)$$

$$= 0.4743 + 0.4105 + 0.4105 + 0.3321 + 0.2160$$

$$= 1.84 \text{ bits/symbol}$$

$$H(S) = 1.84 \text{ bits/symbol}$$

$$\eta = \frac{H(S)}{\overline{L}}$$

$$= \frac{1.84}{1.9} = 0.9684$$

$$\eta = 96.84\%$$

5. A source emits seven symbols with probabilities of 0.25, 0.25, 0.125, 0.125, 0.125, 0.0625, and 0.0626. Compute the Shannon Fano code for this source and calculate the efficiency.

Solution:

Step 1:

Arrange the probabilities in the descending order
0.25, 0.25, 0.125, 0.125, 0.125, 0.0625, 0.0625

Step 2:

Construction of Shannon Fano code:

0.25	0	0		
0.25	0	1		
0.125	1	0	0	
0.125	1	0	1	
0.125	1	1	0	
0.0625	1	1	1	0
0.0625	1	1	1	1

Step 3:

Symbol	Probability	Codeword	Length (l)
S_0	0.25	0 0	2
S_1	0.25	0 1	2
S_2	0.125	1 0 0	3
S_3	0.125	1 0 1	3
S_4	0.125	1 1 0	3
S_5	0.0625	1 1 1 0	4
S_6	0.0625	1 1 1 1	4

Step 4:

To find out efficiency (η), we must calculate the average codeword length ($\overline{L}$) and entropy $H(S)$.

$$\eta = \frac{H(S)}{\overline{L}}$$

where $\overline{L} = \sum_{k=0}^{6} p_k l_k$

$$= p_0 l_0 + p_1 l_1 + p_2 l_2 + p_3 l_3 + p_4 l_4 + p_5 l_5 + p_6 l_6$$

$$= (0.25)(2) + (0.25)(2) + (0.125)(3) + (0.125)(3) + (0.125)(3)$$

$$+ (0.0625)(4) + (0.0625)(4) = 0.5 + 0.5 + 0.375 + 0.375 + 0.375$$

$$+ 0.25 + 0.25$$

$$\overline{L} = 2.625 \text{ bits/symbol}$$

$$H(S) = \sum_{k=0}^{6} p_k \log_2 \left(\frac{1}{p_k} \right)$$

$$= p_0 \log_2 \left(\frac{1}{p_0} \right) + p_1 \log_2 \left(\frac{1}{p_1} \right) + p_2 \log_2 \left(\frac{1}{p_2} \right) + p_3 \log_2 \left(\frac{1}{p_3} \right)$$

$$+ p_4 \log_2 \left(\frac{1}{p_4} \right) + p_5 \log_2 \left(\frac{1}{p_5} \right) + p_6 \log_2 \left(\frac{1}{p_6} \right)$$

$$= 0.25 \log_2 \left(\frac{1}{0.25} \right) + 0.25 \log_2 \left(\frac{1}{0.25} \right) + 0.125 \log_2 \left(\frac{1}{0.125} \right)$$

$$+ 0.125 \log_2 \left(\frac{1}{0.125} \right) + 0.125 \log_2 \left(\frac{1}{0.125} \right) + 0.0625 \log_2 \left(\frac{1}{0.0625} \right)$$

$$+ 0.0625 \log_2 \left(\frac{1}{0.0625} \right)$$

$$= 0.25 \frac{\log_{10} \left(\frac{1}{0.25} \right)}{\log_{10} 2} + 0.25 \frac{\log_{10} \left(\frac{1}{0.25} \right)}{\log_{10} 2} + 0.125 \frac{\log_{10} \left(\frac{1}{0.125} \right)}{\log_{10} 2}$$

$$+ 0.125 \frac{\log_{10} \left(\frac{1}{0.125} \right)}{\log_{10} 2} + 0.125 \frac{\log_{10} \left(\frac{1}{0.125} \right)}{\log_{10} 2} + 0.0625 \frac{\log_{10} \left(\frac{1}{0.0625} \right)}{\log_{10} 2}$$

$$+ 0.0625 \frac{\log_{10} \left(\frac{1}{0.0625} \right)}{\log_{10} 2}$$

$$= (0.25 \times 2) + (0.25 \times 2) + (0.125 \times 3) + (0.125 \times 3) + (0.125 \times 3)$$

$$+ (0.0625 \times 4) + (0.0625 \times 4)$$

$$= 0.5 + 0.5 + 0.375 + 0.375 + 0.375 + 0.25 + 0.25$$

$$= 2.625 \text{ bits/symbol}$$

$$H(S) = 2.625 \text{ bits/symbol}$$

$$\eta = \frac{H(S)}{\overline{L}}$$

$$= \frac{2.625}{2.625} = 1$$

$$\eta = 100\%$$

6. A discrete memoryless source has an alphabet of seven numbers whose probability of occurrence is as described below:

Messages	X_0	X_1	X_2	X_3	X_4	X_5	X_6
Probability	$\frac{1}{4}$	$\frac{1}{8}$	$\frac{1}{16}$	$\frac{1}{16}$	$\frac{1}{8}$	$\frac{1}{4}$	$\frac{1}{8}$

Solution:

Step 1:

Arrange the probabilities in the descending order

$$\frac{1}{4}, \frac{1}{4}, \frac{1}{8}, \frac{1}{8}, \frac{1}{8}, \frac{1}{16}, \frac{1}{16}$$

Step 2:

Construction of Shannon Fano code:

$\frac{1}{4}$	0	0		
$\frac{1}{4}$	0	1		
$\frac{1}{8}$	1	0	0	
$\frac{1}{8}$	1	0	1	
$\frac{1}{8}$	1	1	0	
$\frac{1}{16}$	1	1	1	0
$\frac{1}{16}$	1	1	1	1

Step 3:

Symbol	Probability	Codeword	Length (l)
X_0	$1/4$	0 0	2
X_1	$1/4$	0 1	2
X_2	$1/8$	1 0 0	3
X_3	$1/8$	1 0 1	3
X_4	$1/8$	1 1 0	3
X_5	$1/16$	1 1 1 0	4
X_6	$1/16$	1 1 1 1	4

Step 4:
To find out efficiency (η), we must calculate the average codeword length $(\overline{L})$ and entropy $H(S)$.

$$\eta = \frac{H(S)}{\overline{L}}$$

where $\overline{L} = \sum_{k=0}^{6} p_k l_k$

$$= p_0 l_0 + p_1 l_1 + p_2 l_2 + p_3 l_3 + p_4 l_4 + p_5 l_5 + p_6 l_6$$

$$= \frac{1}{4}(2) + \frac{1}{4}(2) + \frac{1}{8}(3) + \frac{1}{8}(3) + \frac{1}{8}(3) + \frac{1}{16}(4) + \frac{1}{16}(4)$$

$$= 2.625 \text{ bits/symbol}$$

$$\overline{L} = 2.625 \text{ bits/symbol}$$

$$H(S) = \sum_{k=0}^{6} p_k \log_2 \left(1/p_k \right)$$

$$= p_0 \log_2 \left(\frac{1}{p_0} \right) + p_1 \log_2 \left(\frac{1}{p_1} \right) + p_2 \log_2 \left(\frac{1}{p_2} \right) + p_3 \log_2 \left(\frac{1}{p_3} \right)$$
$$+ p_4 \log_2 \left(\frac{1}{p_4} \right) + p_5 \log_2 \left(\frac{1}{p_5} \right) + p_6 \log_2 \left(\frac{1}{p_6} \right)$$

$$= \frac{1}{4} \frac{\log_{10}\left(\frac{1}{1/4}\right)}{\log_{10} 2} + \frac{1}{4} \frac{\log_{10}\left(\frac{1}{1/4}\right)}{\log_{10} 2} + \frac{1}{8} \frac{\log_{10}\left(\frac{1}{1/8}\right)}{\log_{10} 2} + \frac{1}{8} \frac{\log_{10}\left(\frac{1}{1/8}\right)}{\log_{10} 2}$$

$$+ \frac{1}{8} \frac{\log_{10}\left(\frac{1}{1/8}\right)}{\log_{10} 2} + \frac{1}{16} \frac{\log_{10}\left(\frac{1}{1/16}\right)}{\log_{10} 2} + \frac{1}{16} \frac{\log_{10}\left(\frac{1}{1/16}\right)}{\log_{10} 2}$$

$$= 2.625 \text{ bits/symbol}$$

$$H\left(S\right) = 2.625 \text{ bits/symbol}$$

$$\eta = \frac{H\left(S\right)}{\overline{L}}$$

$$= \frac{2.625}{2.625} = 1$$

$$\eta = 100\%$$

7. Consider a discrete memoryless source S, it emits eight possible messages $s_0, s_1, s_2, s_3, s_4, s_5, s_6, s_7$ with their probabilities

Messages	s_0	s_1	s_2	s_3	s_4	s_5	s_6	s_7
Probability	$1/2$	$1/8$	$1/8$	$1/16$	$1/16$	$1/16$	$1/32$	$1/32$

Solution:
Step 1:
   Arrange the probabilities in the descending order

$$1/2, 1/8, 1/8, 1/16, 1/16, 1/16, 1/32, 1/32$$

Step 2:
   Construction of Shannon Fano code:

$$\begin{array}{c|ccccc} \frac{1}{2} & 0 \\ \hline \frac{1}{8} & 1 & 0 & 0 \\ \hline \frac{1}{8} & 1 & 0 & 1 \\ \hline \frac{1}{16} & 1 & 1 & 0 & 0 \\ \hline \frac{1}{16} & 1 & 1 & 0 & 1 \\ \hline \frac{1}{16} & 1 & 1 & 1 & 0 \\ \hline \frac{1}{32} & 1 & 1 & 1 & 1 & 0 \\ \hline \frac{1}{32} & 1 & 1 & 1 & 1 & 1 \end{array}$$

Step 3:

Symbol	Probability	Codeword	Length (l)
s_0	$1/2$	0	1
s_1	$1/8$	1 0 0	3
s_2	$1/8$	1 0 1	3
s_3	$1/16$	1 1 0 0	4
s_4	$1/16$	1 1 0 1	4
s_5	$1/16$	1 1 1 0	4
s_6	$1/32$	1 1 1 1 0	5
s_7	$1/32$	1 1 1 1 1	5

Step 4:

To find out efficiency (η), we must calculate the average codeword length ($\overline{L}$) and entropy $H(S)$.

$$\eta = \frac{H(S)}{\overline{L}}$$

Where $\overline{L} = \sum_{k=0}^{7} p_k l_k$

$$= p_0 l_0 + p_1 l_1 + p_2 l_2 + p_3 l_3 + p_4 l_4 + p_5 l_5 + p_6 l_6 + p_7 l_7$$

$$= \frac{1}{2}(1) + \frac{1}{8}(3) + \frac{1}{8}(3) + \frac{1}{16}(4) + \frac{1}{16}(4) + \frac{1}{16}(4) + \frac{1}{32}(5) + \frac{1}{32}(5)$$

$$= 2.3125 \text{ bits/symbol}$$

$$\overline{L} = 2.3125 \text{ bits/symbol}$$

$$H\left(S\right) = \sum_{k=0}^{7} p_k \log_2 \left(1/p_k\right)$$

$$= p_0 \log_2 \left(\frac{1}{p_0}\right) + p_1 \log_2 \left(\frac{1}{p_1}\right) + p_2 \log_2 \left(\frac{1}{p_2}\right) + p_3 \log_2 \left(\frac{1}{p_3}\right)$$

$$+ p_4 \log_2 \left(\frac{1}{p_4}\right) + p_5 \log_2 \left(\frac{1}{p_5}\right) + p_6 \log_2 \left(\frac{1}{p_6}\right) + p_7 \log_2 \left(\frac{1}{p_7}\right)$$

$$= \frac{1}{2} \frac{\log_{10}\left(\frac{1}{1/2}\right)}{\log_{10} 2} + \frac{1}{8} \frac{\log_{10}\left(\frac{1}{1/8}\right)}{\log_{10} 2} + \frac{1}{8} \frac{\log_{10}\left(\frac{1}{1/8}\right)}{\log_{10} 2}$$

$$+ \frac{1}{16} \frac{\log_{10}\left(\frac{1}{1/16}\right)}{\log_{10} 2} + \frac{1}{16} \frac{\log_{10}\left(\frac{1}{1/16}\right)}{\log_{10} 2} + \frac{1}{16} \frac{\log_{10}\left(\frac{1}{1/16}\right)}{\log_{10} 2}$$

$$+ \frac{1}{32} \frac{\log_{10}\left(\frac{1}{1/32}\right)}{\log_{10} 2} + \frac{1}{32} \frac{\log_{10}\left(\frac{1}{1/32}\right)}{\log_{10} 2}$$

$$= \frac{1}{2} \frac{\log_{10}(2)}{\log_{10} 2} + \frac{1}{8} \frac{\log_{10}(8)}{\log_{10} 2} + \frac{1}{8} \frac{\log_{10}(8)}{\log_{10} 2} + \frac{1}{16} \frac{\log_{10}(16)}{\log_{10} 2}$$

$$+ \frac{1}{16} \frac{\log_{10}(16)}{\log_{10} 2} + \frac{1}{16} \frac{\log_{10}(16)}{\log_{10} 2} + \frac{1}{32} \frac{\log_{10}(32)}{\log_{10} 2} + \frac{1}{32} \frac{\log_{10}(32)}{\log_{10} 2}$$

$$= \frac{1}{2} + \frac{1}{8}(3) + \frac{1}{8}(3) + \frac{1}{16}(4) + \frac{1}{16}(4) + \frac{1}{16}(4) + \frac{1}{32}(5) + \frac{1}{32}(5)$$

$$= 0.5 + 0.375 + 0.375 + 0.25 + 0.25 + 0.25 + 0.15625 + 0.15625$$

$$= 2.312 \text{ bits/symbol}$$

$$H\left(S\right) = 2.3125 \text{ bits/symbol}$$

$$\eta = \frac{H\left(S\right)}{L}$$

$$= \frac{2.3125}{2.3125} = 1$$

$$\eta = 100\%$$

6.5.2 Solved Problems on Huffman-Coding Algorithm

1. Find out the Huffman code for a discrete memory less source with probability statistics

$$\{0.1, \ 0.1, \ 0.2, \ 0.2, \ 0.4\}.$$

Solution:

Step 1:

  Arrange the probabilities in the descending order
  0.4, 0.2, 0.2, 0.1, 0.1

Step 2:

  (i) At stage 1, assign 0 and 1 to two messages of the lowest probability. Here, the lowest probability of message points are 0.1 and 0.1.

0.4

0.2

0.2

0.1 ——— 0

0.1 ——— 1

Stage 1

  (ii) Combine these lowest probability messages and rearrange the order in stage 2.

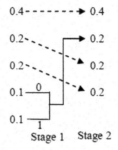

(iii) At stage 2, assign 0 and 1 to two messages of the lowest probability. Here the lowest probabilities of message points are 0.2 and 0.2.

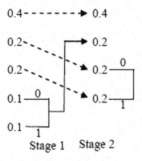

(iv) Combine these lowest probability messages and rearrange the order in stage 3.

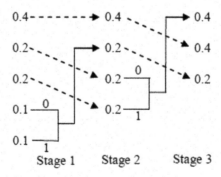

(v) At stage 3, assign 0 and 1 to two messages of the lowest probability. Here, the lowest probability of message points are 0.4 and 0.2.

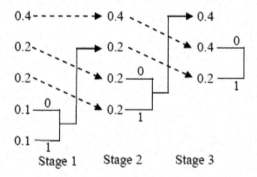

(vi) Combine these lowest probability messages and rearrange the order in stage 4.

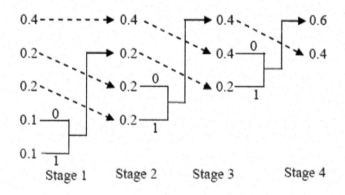

(vii) At stage 4, assign 0 and 1 to two messages of lowest probability. Here, the lowest probability of message points are 0.6 and 0.4.

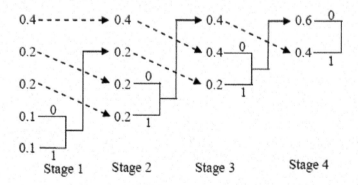

Further splitting of stages is not possible. Hence, stop the iteration.
Step 3:
Codeword calculation:

- Symbol 1; whose probability is 0.4:

0.4 in stage 1 is moved to 0.4 in stage 2 and not assigned any codeword. At stage 3, 0.4 is assigned to codeword 0. At stage 4, 0.4 is modified into 0.6 and codeword for 0.6 is 0 (i.e. combination of 0.4 and 0.2 at stage 3 produces 0.6 at stage 4). Final codewords are obtained from the right direction to the left-most part. Therefore, codeword for symbol 1, whose probability is 0.4 is "0 0".

- Codeword for symbol 2, whose probability is 0.2:

0.2 in stage 1 is moved to 0.2 in stage 2 with codeword 0. At stage 3, 0.2 is moved to 0.2 with codeword 1. At stage 4, 0.2 is modified into 0.6 and codeword for 0.6 is 0. (i.e. combination of 0.4 and 0.2 at stage 3 produces 0.6 at stage 4)

The final codeword for symbol 2 is 1 0. (Written from the right-most direction to the left-most direction.)

- Codeword for symbol 3, whose probability is 0.2:

0.2 in stage 1 is moved to 0.2 in stage 2 and assigned by a codeword of 1. 0.2 in stage 2 is modified into 0.4 in stage 3 and it is not assigned any codeword (i.e. combination of probability 0.2 and 0.2 in stage 2 produces 0.4 in stage 3). 0.4 at stage 3 is moved to 0.4 in stage 4, with the codeword of 1.

Therefore, the final codeword for symbol 3 is 1 1.

- Codeword for symbol 4, whose probability is 0.1:

0.1 in stage 1 is modified into 0.2 in stage 2 (i.e. combination of probabilities 0.1 and 0.1 in stage 1 produces 0.2 in stage 2, and it is not assigned any codeword). 0.2 in stage 2 is moved to 0.2 in stage 3 and assigned by a codeword of 1. 0.2 in stage 3 is modified into 0.6 and assigned by a codeword of 0 (i.e. combination of probabilities 0.4 and 0.2 in stage 3 produces 0.6 in stage 4).

Final codeword for symbol 4 is 0 1 0.

- Codeword for symbol 5, whose probability is 0.1:

0.1 in stage 1 is modified into 0.2 in stage 2 and it is not assigned any codeword (i.e. combination of 0.1 and 0.1 in stage 1 produces 0.2 at stage 2). 0.2 in stage 2 is moved to 0.2 in stage 3, with the codeword of 1. 0.2 at stage 3, modified into 0.6 at stage 4, with the codeword of 0 (i.e. combination of probabilities 0.4 and 0.2 at stage 3 produces 0.6 at stage 4).

Final codeword for symbol 5 is 0 1 1.

Step 4:

Symbol	Probability	Codeword	Length (l)
S_0	0.4	0 0	2
S_1	0.2	10	2
S_2	0.2	1 1	2
S_3	0.1	0 1 0	3
S_4	0.1	0 1 1	3

Step 5:

To find out efficiency (η), we must calculate the average codeword length ($\overline{L}$) and entropy $H(S)$.

$$\eta = \frac{H(S)}{\overline{L}}$$

where $\overline{L} = \sum_{k=0}^{4} p_k l_k$

$$= p_0 l_0 + p_1 l_1 + p_2 l_2 + p_3 l_3 + p_4 l_4$$

$$= (0.4)(2) + (0.2)(2) + (0.2)(2) + (0.1)(3) + (0.1)(3)$$

$$= 0.8 + 0.4 + 0.4 + 0.3 + 0.3$$

$$\overline{L} = 2.2 \text{ bits/symbol}$$

$$H(S) = \sum_{k=0}^{4} p_k \log_2 \left(1/p_k \right)$$

$$= p_0 \log_2 \left(\frac{1}{p_0} \right) + p_1 \log_2 \left(\frac{1}{p_1} \right) + p_2 \log_2 \left(\frac{1}{p_2} \right) + p_3 \log_2 \left(\frac{1}{p_3} \right)$$

$$+ p_4 \log_2 \left(\frac{1}{p_4} \right)$$

$$= 0.4 \log_2 \left(\frac{1}{0.4}\right) + 0.2 \log_2 \left(\frac{1}{0.2}\right) + 0.2 \log_2 \left(\frac{1}{0.2}\right) + 0.1 \log_2 \left(\frac{1}{0.1}\right)$$

$$+ 0.1 \log_2 \left(\frac{1}{0.1}\right)$$

$$= 0.4 \frac{\log_{10} \left(\frac{1}{0.4}\right)}{\log_{10} 2} + 0.2 \frac{\log_{10} \left(\frac{1}{0.2}\right)}{\log_{10} 2} + 0.2 \frac{\log_{10} \left(\frac{1}{0.2}\right)}{\log_{10} 2} + 0.1 \frac{\log_{10} \left(\frac{1}{0.1}\right)}{\log_{10} 2}$$

$$+ 0.1 \frac{\log_{10} \left(\frac{1}{0.1}\right)}{\log_{10} 2}$$

$$= (0.4 \times 1.3219) + (0.2 \times 2.3219) + (0.2 \times 2.3219) + (0.1 \times 3.3219)$$

$$+ (0.1 \times 3.3219)$$

$$= 0.5287 + 0.4643 + 0.4643 + 0.33219 + 0.33219$$

$$= 2.12 \text{ bits/symbol}$$

$$H(S) = 2.12 \text{ bits/symbol}$$

$$\eta = \frac{H(S)}{\overline{L}}$$

$$= \frac{2.12}{2.2} = 0.96$$

$$\eta = 96\%$$

2. A discrete memoryless source has five messages S_1, S_2, S_3, S_4 and S_5 with probabilities

$$p(S_1) = 0.4, p(S_2) = 0.19, p(S_3) = 0.16, p(S_4) = 0.15, p(S_5) = 0.1.$$

Construct the Huffman code and calculate the code efficiency.

Solution:
Step 1:
    Arrange the probabilities in the descending order
    0.4, 0.19, 0.16, 0.15, 0.1

Step 2:

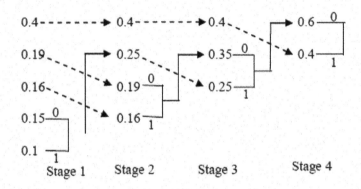

Step 3:
Codewords:

$0.4 \to 1$; $0.19 \to 0\,0\,0$; $0.16 \to 0\,0\,1$; $0.15 \to 0\,1\,0$; $0.1 \to 0\,1\,1$

Step 4:

Symbol	Probability	Codeword	Length (l)
S_1	0.4	1	1
S_2	0.19	0 0 0	3
S_3	0.16	0 0 1	3
S_4	0.15	0 1 0	3
S_5	0.1	0 1 1	3

Step 5:
    To find out efficiency (η), we must calculate the average codeword length ($\overline{L}$) and entropy $H(S)$.

$$\eta = \frac{H(S)}{\overline{L}}$$

where $\overline{L} = \sum_{k=0}^{4} p_k l_k$

$$= p_0 l_0 + p_1 l_1 + p_2 l_2 + p_3 l_3 + p_4 l_4$$

$$= (0.4)(1) + (0.19)(3) + (0.16)(3) + (0.15)(3) + (0.1)(3)$$

$$= 0.4 + 0.57 + 0.48 + 0.45 + 0.3$$

$$\overline{L} = 2.2 \text{ bits/symbol}$$

$$H(S) = \sum_{k=1}^{5} p_k \log_2 \left(\frac{1}{p_k} \right)$$

$$= p_1 \log_2 \left(\frac{1}{p_1} \right) + p_2 \log_2 \left(\frac{1}{p_2} \right) + p_3 \log_2 \left(\frac{1}{p_3} \right) + p_5 \log_2 \left(\frac{1}{p_5} \right)$$

$$= 0.4 \log_2 \left(\frac{1}{0.4} \right) + 0.19 \log_2 \left(\frac{1}{0.2} \right) + 0.16 \log_2 \left(\frac{1}{0.2} \right)$$

$$+ 0.15 \log_2 \left(\frac{1}{0.1} \right) + 0.1 \log_2 \left(\frac{1}{0.1} \right)$$

$$= 0.4 \frac{\log_{10} \left(\frac{1}{0.4} \right)}{\log_{10} 2} + 0.19 \frac{\log_{10} \left(\frac{1}{0.19} \right)}{\log_{10} 2} + 0.16 \frac{\log_{10} \left(\frac{1}{0.16} \right)}{\log_{10} 2} + 0.15 \frac{\log_{10} \left(\frac{1}{0.15} \right)}{\log_{10} 2}$$

$$+ 0.1 \frac{\log_{10} \left(\frac{1}{0.1} \right)}{\log_{10} 2}$$

$$= (0.4 \times 1.3219) + (0.19 \times 2.3959) + (0.16 \times 2.6438) + (0.15 \times 2.7369)$$

$$+ (0.1 \times 3.3219)$$

$$= 0.5287 + 0.4552 + 0.4230 + 0.4105 + 0.3321$$

$$= 2.14 \text{ bits/symbol}$$

$$H(S) = 2.14 \text{ bits/symbol}$$

$$\eta = \frac{H(S)}{\overline{L}}$$

$$= \frac{2.14}{2.2} = 0.9727$$

$$\eta = 97.27\%$$

3) A discrete memoryless source emits six symbols S_0, S_1, S_2, S_3, S_4 and S_5 with probabilities 0.3, 0.25, 0.05, 0.12, 0.08, and 0.2, respectively. Construct the Huffman code and compute its efficiency.

Solution:

Step 1:

Arrange the probabilities in the descending order
0.3, 0.25, 0.2, 0.12, 0.08, 0.05

Step 2: Construction of Huffman code

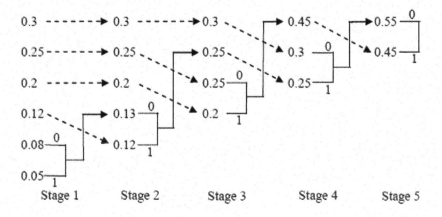

Step 3: Codeword calculation

$0.3 \rightarrow 0\ 0;\ 0.25 \rightarrow 1\ 0;\ 0.2 \rightarrow 1\ 1;\ 0.12 \rightarrow 0\ 1\ 1;\ 0.08 \rightarrow 0\ 1\ 0\ 0; 0.05$

$$\rightarrow 0\ 1\ 0\ 1;$$

Step 4:

Symbol	Probability	Codeword	Length (l)
S_0	0.3	0 0	2
S_1	0.25	0 1	2
S_2	0.2	1 1	2
S_3	0.12	0 1 0	3
S_4	0.08	0 1 0 0	4
S_5	0.05	0 1 0 1	4

Step 5:

To find out efficiency (η), we must calculate the average codeword length $(\overline{L})$ and entropy $H(S)$.

$$\eta = \frac{H(S)}{\overline{L}}$$

where $\overline{L} = \sum_{k=0}^{5} p_k l_k$

$$= p_0 l_0 + p_1 l_1 + p_2 l_2 + p_3 l_3 + p_4 l_4 + p_5 l_5$$

$$= (0.3)(2) + (0.25)(2) + (0.2)(2) + (0.12)(3) + (0.08)(4) + (0.05)(4)$$

$$= 0.6 + 0.5 + 0.4 + 0.36 + 0.32 + 0.2$$

$$\overline{L} = 2.38 \text{ bits/symbol}$$

$$H(S) = \sum_{k=0}^{5} p_k \log_2 \left(\frac{1}{p_k}\right)$$

$$= p_0 \log_2 \left(\frac{1}{p_0}\right) + p_1 \log_2 \left(\frac{1}{p_1}\right) + p_2 \log_2 \left(\frac{1}{p_2}\right) + p_3 \log_2 \left(\frac{1}{p_3}\right)$$

$$+ p_4 \log_2 \left(\frac{1}{p_4}\right) + p_5 \log_2 \left(\frac{1}{p_5}\right)$$

$$= 0.3 \log_2 \left(\frac{1}{0.3}\right) + 0.25 \log_2 \left(\frac{1}{0.25}\right) + 0.2 \log_2 \left(\frac{1}{0.2}\right)$$

$$+ 0.12 \log_2 \left(\frac{1}{0.12}\right) + 0.08 \log_2 \left(\frac{1}{0.08}\right) + 0.05 \log_2 \left(\frac{1}{0.05}\right)$$

$$= 0.3 \frac{\log_{10} \left(\frac{1}{0.3}\right)}{\log_{10} 2} + 0.25 \frac{\log_{10} \left(\frac{1}{0.25}\right)}{\log_{10} 2} + 0.2 \frac{\log_{10} \left(\frac{1}{0.2}\right)}{\log_{10} 2} + 0.12 \frac{\log_{10} \left(\frac{1}{0.12}\right)}{\log_{10} 2}$$

$$+ 0.08 \frac{\log_{10} \left(\frac{1}{0.08}\right)}{\log_{10} 2} + 0.05 \frac{\log_{10} \left(\frac{1}{0.05}\right)}{\log_{10} 2}$$

$$= (0.3 \times 1.7369) + (0.25 \times 2) + (0.2 \times 2.3219) + (0.12 \times 3.0588)$$

$$+ (0.08 \times 3.6438) + (0.05 \times 4.3219)$$

$$= 0.5210 + 0.5 + 0.4643 + 0.3670 + 0.2915 + 0.2160$$

$$= 2.35 \text{ bits/symbol}$$

$$H(S) = 2.35 \text{ bits/symbol}$$

$$\eta = \frac{H(S)}{\overline{L}}$$

$$= \frac{2.35}{2.38} = 0.9873$$

$$\eta = 98.73\%$$

4. A discrete memoryless source emits five symbols with probabilities 0.15, 0.10, 0.05, 0.15, and 0.55, respectively. Compute the efficiency using Huffman coding algorithm.

Solution:

Step 1:

Arrange the probabilities in the decreasing order

0.55, 0.15, 0.15, 0.10, 0.05

Step 2: Construction of Huffman code

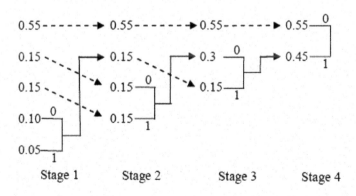

Step 3: Codeword calculation

$0.55 \rightarrow 0 ; 0.15 \rightarrow 1\,0\,0 ; 0.15 \rightarrow 1\,0\,1 ; 0.10 \rightarrow 1\,1\,0 ; 0.05 \rightarrow 1\,1\,1$

Step 4:

Symbol	Probability	Codeword	Length (l)
S_0	0.55	0	1
S_1	0.15	1 0 0	3
S_2	0.15	1 0 1	3
S_3	0.10	1 1 0	3
S_4	0.05	1 1 1	3

Step 5:

To find out efficiency (η), we must calculate the average codeword length ($\overline{L}$) and entropy $H(S)$.

$$\eta = \frac{H(S)}{\overline{L}}$$

where $\overline{L} = \sum_{k=0}^{4} p_k l_k$

$$= p_0 l_0 + p_1 l_1 + p_2 l_2 + p_3 l_3 + p_4 l_4$$

$$= (0.55)(1) + (0.15)(3) + (0.15)(3) + (0.1)(3) + (0.05)(3)$$

$$= 0.55 + 0.45 + 0.45 + 0.3 + 0.15$$

$$\overline{L} = 1.9 \text{ bits/symbol}$$

$$H(S) = \sum_{k=0}^{4} p_k \log_2 \left(1/p_k\right)$$

$$= p_0 \log_2 \left(\frac{1}{p_0}\right) + p_1 \log_2 \left(\frac{1}{p_1}\right) + p_2 \log_2 \left(\frac{1}{p_2}\right) + p_3 \log_2 \left(\frac{1}{p_3}\right)$$

$$+ p_4 \log_2 \left(\frac{1}{p_4}\right)$$

$$= 0.55 \log_2\left(\frac{1}{0.55}\right) + 0.15 \log_2\left(\frac{1}{0.15}\right) + 0.15 \log_2\left(\frac{1}{0.15}\right)$$

$$+ 0.1 \log_2\left(\frac{1}{0.1}\right) + 0.05 \log_2\left(\frac{1}{0.05}\right)$$

$$= 0.55\frac{\log_{10}\left(\frac{1}{0.55}\right)}{\log_{10} 2} + 0.15\frac{\log_{10}\left(\frac{1}{0.15}\right)}{\log_{10} 2} + 0.15\frac{\log_{10}\left(\frac{1}{0.15}\right)}{\log_{10} 2}$$

$$+ 0.1\frac{\log_{10}\left(\frac{1}{0.1}\right)}{\log_{10} 2} + 0.05\frac{\log_{10}\left(\frac{1}{0.05}\right)}{\log_{10} 2}$$

$$= (0.55 \times 0.8624) + (0.15 \times 2.7369) + (0.15 \times 2.7369)$$

$$+ (0.10 \times 3.3219) + (0.05 \times 4.3219)$$

$$= 0.4743 + 0.4105 + 0.4105 + 0.3321 + 0.2160$$

$$= 1.84 \text{ bits/symbol}$$

$$H(S) = 1.84 \text{ bits/symbol}$$

$$\eta = \frac{H(S)}{\overline{L}}$$

$$= \frac{1.84}{1.9} = 0.9684$$

$$\eta = 96.84\%$$

5. A source emits seven symbols with the probabilities of 0.25, 0.25, 0.125, 0.125, 0.125, 0.0625, and 0.0626. Compute the Huffman code for this source and calculate the efficiency.

Solution:

Step 1:

Arrange the probabilities in the decreasing order
0.25, 0.25, 0.125, 0.125, 0.125, 0.0625, 0.0625

Step 2:

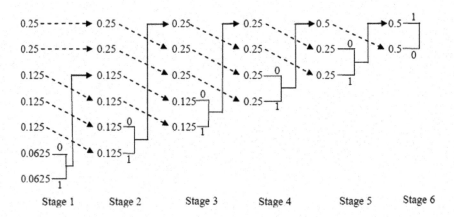

Stage 1	Stage 2	Stage 3	Stage 4	Stage 5	Stage 6

Step 3: Codeword calculation

$0.25 \rightarrow 1\,0$; $0.25 \rightarrow 1\,1$; $0.125 \rightarrow 0\,0\,1$; $0.125 \rightarrow 0\,1\,0$; $0.125 \rightarrow 0\,1\,1$;

$$0.0625 \rightarrow 0\,0\,0\,0; 0.0625 \rightarrow 0\,0\,0\,1$$

Step 4:

Symbol	Probability	Codeword	Length
S_0	0.25	1 0	2
S_1	0.25	1 1	2
S_2	0.125	0 0 1	3
S_3	0.125	0 1 0	3
S_4	0.125	0 1 1	3
S_5	0.0625	0 0 0 0	4
S_6	0.0625	0 0 0 1	4

Step 5:

To find out efficiency (η), we must calculate the average codeword length $(\overline{L})$ and entropy $H(S)$.

$$\eta = \frac{H(S)}{\overline{L}}$$

where $\overline{L} = \sum_{k=0}^{6} p_k l_k$

$$= p_0 l_0 + p_1 l_1 + p_2 l_2 + p_3 l_3 + p_4 l_4 + p_5 l_5 + p_6 l_6$$

$$= (0.25)\,(2) + (0.25)\,(2) + (0.125)\,(3) + (0.125)\,(3) + (0.125)\,(3)$$
$$+ (0.0625)\,(4) + (0.0625)\,(4)$$
$$= 0.5 + 0.5 + 0.375 + 0.375 + 0.375 + 0.25 + 0.25$$

$$\overline{L} = 2.625 \text{ bits/symbol}$$

$$H\,(S) = \sum_{k=0}^{6} p_k \log_2\left(1/p_k\right)$$

$$= p_0 \log_2\left(\frac{1}{p_0}\right) + p_1 \log_2\left(\frac{1}{p_1}\right) + p_2 \log_2\left(\frac{1}{p_2}\right) + p_3 \log_2\left(\frac{1}{p_3}\right)$$
$$+ p_4 \log_2\left(\frac{1}{p_4}\right) + p_5 \log_2\left(\frac{1}{p_5}\right) + p_6 \log_2\left(\frac{1}{p_6}\right)$$

$$= 0.25 \log_2\left(\frac{1}{0.25}\right) + 0.25 \log_2\left(\frac{1}{0.25}\right) + 0.125 \log_2\left(\frac{1}{0.125}\right)$$
$$+ 0.125 \log_2\left(\frac{1}{0.125}\right) + 0.125 \log_2\left(\frac{1}{0.125}\right)$$
$$+ 0.0625 \log_2\left(\frac{1}{0.0625}\right) + 0.0625 \log_2\left(\frac{1}{0.0625}\right)$$

$$= 0.25\frac{\log_{10}\left(\frac{1}{0.25}\right)}{\log_{10} 2} + 0.25\frac{\log_{10}\left(\frac{1}{0.25}\right)}{\log_{10} 2} + 0.125\frac{\log_{10}\left(\frac{1}{0.125}\right)}{\log_{10} 2}$$
$$+ 0.125\frac{\log_{10}\left(\frac{1}{0.125}\right)}{\log_{10} 2} + 0.125\frac{\log_{10}\left(\frac{1}{0.125}\right)}{\log_{10} 2} + 0.0625\frac{\log_{10}\left(\frac{1}{0.0625}\right)}{\log_{10} 2}$$
$$+ 0.0625\frac{\log_{10}\left(\frac{1}{0.0625}\right)}{\log_{10} 2}$$

$$= (0.25 \times 2) + (0.25 \times 2) + (0.125 \times 3) + (0.125 \times 3) + (0.125 \times 3)$$
$$+ (0.0625 \times 4) + (0.0625 \times 4)$$

$$= 0.5 + 0.5 + 0.375 + 0.375 + 0.375 + 0.25 + 0.25$$

$$= 2.625 \text{ bits/symbol}$$

$$H\left(S\right) = 2.625 \text{ bits/symbol}$$

$$\eta = \frac{H\left(S\right)}{\overline{L}}$$

$$= \frac{2.625}{2.625} = 1$$

$$\eta = 100\%$$

7

Channel Coding

7.1 Shannon's Theorems

7.1.1 Shannon's First Theorem: Source Coding Theorem

Coding efficiency is given by

$$\eta = \frac{L_{\min}}{\overline{L}} \tag{7.1}$$

where $\overline{L}$ represents the average number of bits per symbol and called code length, and $L_{\min}$ denotes the minimum possible value of $\overline{L}$.

Given a discrete memoryless source of entropy $H(S)$, the average codeword length $\overline{L}$ for any source coding is bounded as

$$\overline{L} \geq H(S) \tag{7.2}$$

Here, $H(S)$ represents the fundamental limit on the average number of bits per symbol. Hence, we can make $H(S) \approx L_{\min}$; therefore, efficiency is given by

$$\eta = \frac{H(S)}{\overline{L}} \tag{7.3}$$

7.1.2 Shannon's Second Theorem: Channel Coding Theorem

Consider a band-limited channel, whose bandwidth is 'B', and which carries a signal having M number of levels. Then, the maximum data rate 'R' of the channel is

$$R = 2B \log_2 M \tag{7.4}$$

For noiseless channel data rate, R should satisfy the following condition:

$R \ll C$. This condition is referred to as channel coding theorem or Shannon's second theorem. Here, C represents the capacity of the channel.

7.1.3 Shannon's Third Theorem: Channel Capacity Theorem or Shannon's Hartley Theorem

The channel capacity of a white, band-limited Gaussian channel is given by

$$C = B \log_2 \left(1 + {}^S/_N\right) \text{ bits/sec} \qquad (7.5)$$

where B is the channel bandwidth, S is the signal power, and N is the noise within the channel bandwidth and it is referred to as noise power.

$$N = N_O B.$$

$(N_O/2)$ is the power spectral density of white noise.

Capacity C depends on two factors B and S/N ratio. We have to find out the maximum possible value of C.

- **Effect of S/N on C:**

Let us assume the communication channel as noiseless, then $N = 0$, therefore $S/N \to \infty$. Hence, capacity C also tends to ∞. Thus, a noiseless channel has *infinite capacity*.

- **Effect of bandwidth (B) on channel capacity:**

Let us consider some noise is present in the channel. Then S/N is not infinite.

Now if the bandwidth approaches to ∞ (infinity), the channel capacity does not become infinite, since $N = N_O B$, i.e. noise power also increases with the channel bandwidth B. This reduces the value of (S/N), with increase in B, assuming signal power (S) as constant.

Therefore, we can conclude that a channel with an infinite bandwidth has a finite channel capacity. It is denoted by C_∞.

$$C_\infty = 1.44 S/N_O$$

Proof:

Channel with infinite bandwidth has finite capacity:

Channel capacity C is given by

$$C = B \log_2 \left(1 + {}^S/_N\right) \qquad (7.6)$$

Noise power N is given by,

$$N = N_O B \qquad (7.7)$$

Substitute Equation (7.7) in (7.6),

$$C = B \log_2 \left(1 + \frac{S}{N_O B}\right)$$

$$= \left(\frac{S}{N_O}\right) \left(\frac{N_O}{S}\right) B \log_2 \left(1 + \frac{S}{N_O B}\right)$$

$$= \left(\frac{S}{N_O}\right) \left(\frac{N_O B}{S}\right) \log_2 \left(1 + \frac{S}{N_O B}\right)$$

$$= \left(\frac{S}{N_O}\right) \log_2 \left[\left(1 + \frac{S}{N_O B}\right)\right]^{\frac{N_O B}{S}}$$

When $B \to \infty$;

$$C_\infty = \left(\frac{S}{N_O}\right) \log_2 e \qquad \because \underset{x \to 0}{Lt} (1+x)^{\left(\frac{1}{x}\right)} = e$$

$$C_\infty = \left(\frac{S}{N_O}\right) 1.44 \tag{7.8}$$

Summary:

Channel capacity C depends on bandwidth B and *S/N*.

- Noiseless channel has infinite capacity.
- Channel with infinite bandwidth has a finite channel capacity.

This theorem is applicable to Gaussian noise channel.

7.1.4 Mutual Information

It gives the uncertainty of the channel input that is resolved by observing the channel output. Mathematically it can be expressed as:

$$I(X,Y) = H(X) - H\left(\frac{X}{Y}\right) \tag{7.9}$$

where $H(X)$ represents entropy of the channel input, $H(X/Y)$ is conditional entropy of channel input that is resolved by observing the channel output Y.

Figure 7.1 shows the mutual information.

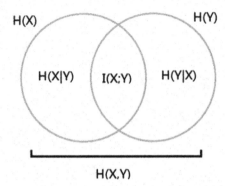

H(X,Y)

Figure 7.1   Mutual information.

7.2 Error Control Techniques: Linear Block Codes

Types of error control coding approaches:

  i. Linear block codes
 ii. Cyclic codes and
iii. Convolutional codes

7.2.1 Linear Block Codes

Linear block codes are represented by the notation (n, k). The channel encoder accepts k message bits and produces the output in the size of n encoded bits. It is illustrated in Figure 7.2. The output n bits consist of k information bits and $(n{-}k)$ parity or redundant bits. For example, $(6, 3)$ linear block code represents

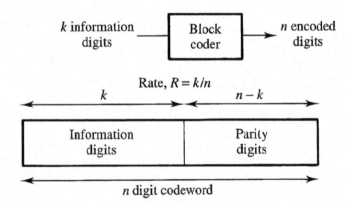

Figure 7.2   Linear block codes.

that input to the encoder is 3 bits $(k = 3)$ and the output of the encoder is 6 bits $(n = 6)$. Here, the redundant or parity bits are 3 bits, since $(n - k) = 6 - 3 = 3$ bits.

Output of the encoder n bits is mathematically represented as

$$[X] = [M] \bullet [G]$$

where X is a codeword matrix of $1 \times n$ size M represents the message codeword of size $1 \times k$ size and G is a generator matrix of $1 \times n$ size.

Generator matrix G can be represented as

$$G = [I_k | P]_{k \times n}$$

where I is $(k \times k)$ identity matrix and P is $k \times (n - k)$ coefficient matrix.

Hence, we need message bits and generator matrix to generate encoded output at the transmitter side. The same generator matrix is used at the receiver side in the name of parity check matrix to decode the message bits.

In summary, a family of (n, k) linear block codes have the following parameters:

1. Number of parity bits: $m = n - k$
2. Number of message bits: $k = 2^m - m - 1$
3. Block length: $n = 2^m - 1$

when $m \geq 3$; these are called *Hamming codes*.

Steps involved in Generation of codewords (Transmitter side):

1. From the given generator matrix, separate out the coefficient matrix P.
2. Generate the redundant or parity bits for each message word by

$$R = M \bullet P$$

3. Combine these parity bits after the message bits to get the corresponding codewords.

7.2.2 Solved Problems on Linear Block Codes

1. Consider a (6, 3) linear block code, whose generator matrix is given below. Find all the codewords.
2.

$$G = \begin{bmatrix} 1 & 0 & 0 & \vdots & 0 & 1 & 1 \\ 0 & 1 & 0 & \vdots & 1 & 0 & 1 \\ 0 & 0 & 1 & \vdots & 1 & 1 & 0 \end{bmatrix}$$

Solution:

The general format of linear block code is (n, k). It is compared with given problems (6, 3) linear block code. It gives, $n = 6$; and $k = 3$. It represents the total number of bits in message bits = 3. (i.e. $k = 3$) and the total number of bits in the codeword is 6. (i.e. $n = 6$). Hence $(n - k)$ is the parity bits (i.e. $6 - 3 = 3$ bits are redundant bits).

Step 1:

Separate the coefficient matrix P from the given generator matrix G.
It is known that

$$G = \left[I_k \vdots P \right].$$

Comparing the above expression with the given generator matrix G, we find that,

Identity matrix $I_k = \begin{bmatrix} 1 & 0 & 0 \\ 0 & 1 & 0 \\ 0 & 0 & 1 \end{bmatrix}$ and coefficient matrix $P = \begin{bmatrix} 0 & 1 & 1 \\ 1 & 0 & 1 \\ 1 & 1 & 0 \end{bmatrix}$

Step 2: Finding parity or redundant bits

It can be calculated using the following formulae.

$$R = M \bullet P \qquad\qquad (7.10)$$

Here, R denotes redundant bits, M represents message bits, and P represents parity bits.

Here, the size of the redundant and the message bits are 3-bit long. Therefore, Equation (7.10) can be expressed as

$$\begin{bmatrix} r_0 & r_1 & r_2 \end{bmatrix} = \begin{bmatrix} m_0 & m_1 & m_2 \end{bmatrix} \begin{bmatrix} 0 & 1 & 1 \\ 1 & 0 & 1 \\ 1 & 1 & 0 \end{bmatrix}$$

Therefore,

$r_0 = (m_0 \times 0) \oplus (m_1 \times 1) \oplus (m_2 \times 1) = (m_1 \times 1) \oplus (m_2 \times 1)$

$r_0 = m_1 \oplus m_2$

$r_1 = (m_0 \times 1) \oplus (m_1 \times 0) \oplus (m_2 \times 1)$

$r_1 = m_0 \oplus m_2$

$r_2 = (m_0 \times 1) \oplus (m_1 \times 1) \oplus (m_2 \times 0)$

$r_2 = m_0 \oplus m_1$

- Finding redundant bits, when the message bit is $[m_0 \quad m_1 \quad m_2] = [0 \quad 0 \quad 0]$

$$r_0 = m_1 \oplus m_2 = 0 \oplus 0 = 0$$
$$r_1 = m_0 \oplus m_2 = 0 \oplus 0 = 0$$
$$r_2 = m_0 \oplus m_1 = 0 \oplus 0 = 0$$

Therefore, the final codeword =

m_0	m_1	m_2	r_0	r_1	r_2
0	0	0	0	0	0

| Message | | | Parity bits | | |

- Finding redundant bits, when the message bit is $[m_0 \quad m_1 \quad m_2] = [0 \quad 0 \quad 1]$

$$r_0 = m_1 \oplus m_2 = 0 \oplus 1 = 1$$
$$r_1 = m_0 \oplus m_2 = 0 \oplus 1 = 1$$
$$r_2 = m_0 \oplus m_1 = 0 \oplus 0 = 0$$

Therefore, the final codeword =

m_0	m_1	m_2	r_0	r_1	r_2
0	0	1	1	1	0

| Message | | | Parity bits | | |

- Finding the redundant bits, when the message bit is $[m_0 \quad m_1 \quad m_2] = [0 \quad 1 \quad 0]$

$$r_0 = m_1 \oplus m_2 = 1 \oplus 0 = 1$$
$$r_1 = m_0 \oplus m_2 = 0 \oplus 0 = 0$$
$$r_2 = m_0 \oplus m_1 = 0 \oplus 1 = 1$$

Therefore, the final codeword =

m_0	m_1	m_2	r_0	r_1	r_2
0	1	0	1	0	1

| Message | | | Parity bits | | |

- Finding the redundant bits, when the message bit is $[m_0 \quad m_1 \quad m_2] = [0 \quad 1 \quad 1]$

$$r_0 = m_1 \oplus m_2 = 1 \oplus 1 = 0$$
$$r_1 = m_0 \oplus m_2 = 0 \oplus 1 = 1$$
$$r_2 = m_0 \oplus m_1 = 0 \oplus 1 = 1$$

Therefore, the final codeword =

m_0	m_1	m_2	r_0	r_1	r_2
0	1	1	0	1	1

| Message | | | Parity bits | | |

- Finding the redundant bits, when the message bit is $[m_0 \quad m_1 \quad m_2] = [1 \quad 0 \quad 0]$

$$r_0 = m_1 \oplus m_2 = 1 \oplus 1 = 0$$
$$r_1 = m_0 \oplus m_2 = 0 \oplus 1 = 1$$
$$r_2 = m_0 \oplus m_1 = 1 \oplus 0 = 1$$

Therefore, the final codeword =

m_0	m_1	m_2	r_0	r_1	r_2
1	0	0	0	1	1
Message			Parity bits		

- Finding the redundant bits, when the message bit is $[m_0 \quad m_1 \quad m_2] = [1 \quad 0 \quad 1]$

$$r_0 = m_1 \oplus m_2 = 0 \oplus 1 = 1$$
$$r_1 = m_0 \oplus m_2 = 1 \oplus 1 = 0$$
$$r_2 = m_0 \oplus m_1 = 1 \oplus 0 = 1$$

Therefore, the final codeword =

m_0	m_1	m_2	r_0	r_1	r_2
1	0	1	1	0	1
Message			Parity bits		

- Finding the redundant bits, when the message bit is $[m_0 \quad m_1 \quad m_2] = [1 \quad 1 \quad 0]$

$$r_0 = m_1 \oplus m_2 = 1 \oplus 0 = 1$$
$$r_1 = m_0 \oplus m_2 = 1 \oplus 0 = 1$$
$$r_2 = m_0 \oplus m_1 = 1 \oplus 1 = 0$$

Therefore, the final codeword =

m_0	m_1	m_2	r_0	r_1	r_2
1	1	0	1	1	0
Message			Parity bits		

- Finding the redundant bits, when the message bit is $[m_0 \quad m_1 \quad m_2] = [1 \quad 1 \quad 1]$

$$r_0 = m_1 \oplus m_2 = 1 \oplus 1 = 0$$
$$r_1 = m_0 \oplus m_2 = 1 \oplus 1 = 0$$
$$r_2 = m_0 \oplus m_1 = 1 \oplus 1 = 0$$

Therefore, the final codeword =

m_0	m_1	m_2	r_0	r_1	r_2
1	1	1	0	0	0
Message			Parity bits		

7.2.3 Solved Problems on Hamming Codes

Hamming codes are linear block codes. Hamming code for $m \geq 3$ is defined by the following parameters.

(i) Block length: $n = 2^m - 1$
(ii) Number of message bits: $k = 2^m - m - 1$
(iii) Number of parity bits: $(n - k) = m$; where $m \geq 3$
(iv) Code rate or code efficiency $= \frac{k}{n} = \frac{2^m - m - 1}{2^m - 1}$

1. Consider a (7, 4) linear block code, whose generator matrix is given below. Find all the codewords.

$$G = \begin{bmatrix} 1 & 0 & 0 & 0 & \vdots & 1 & 1 & 1 \\ 0 & 1 & 0 & 0 & \vdots & 1 & 1 & 0 \\ 0 & 0 & 1 & 0 & \vdots & 1 & 0 & 1 \\ 0 & 0 & 0 & 1 & \vdots & 0 & 1 & 1 \end{bmatrix}$$

Solution:
Given (n, k), linear block code is in the form of (7, 4)
It represents codeword length $(n) = 7$
Number of message bits $(k) = 4$
Number of parity bits $m = n - k = 7 - 4 = 3$
Here $m \geq 3$; therefore, the given block code is *"Hamming code"*.
Step 1: Finding the parity coefficient matrix
Given generator matrix is

$$G = \begin{bmatrix} 1 & 0 & 0 & 0 & \vdots & 1 & 1 & 1 \\ 0 & 1 & 0 & 0 & \vdots & 1 & 1 & 0 \\ 0 & 0 & 1 & 0 & \vdots & 1 & 0 & 1 \\ 0 & 0 & 0 & 1 & \vdots & 0 & 1 & 1 \end{bmatrix} \tag{7.11}$$

As we know, the general format of G is given by the expression

$$G = [I_k \vdots P] \tag{7.12}$$

Comparing Equations (7.11) and (7.12), we find that

$$I_k = \begin{bmatrix} 1 & 0 & 0 & 0 \\ 0 & 1 & 0 & 0 \\ 0 & 0 & 1 & 0 \\ 0 & 0 & 0 & 1 \end{bmatrix}$$

and

$$\text{Coefficient matrix } [P] = \begin{bmatrix} 1 & 1 & 1 \\ 1 & 1 & 0 \\ 1 & 0 & 1 \\ 0 & 1 & 1 \end{bmatrix}$$

Step 2: Finding parity bits
These can be calculated using the following formulae.

$$R = M \bullet P \tag{7.13}$$

Here, R denotes redundant bits, M represents message bits, and P represents parity bits.

In the given problem, the number of message bits is $k = 4$ and number of parity bits or redundant bits is $m = 3$

Therefore, Equation (7.33) can be expressed as

$$\begin{bmatrix} r_0 & r_1 & r_2 \end{bmatrix} = \begin{bmatrix} m_0 & m_1 & m_2 & m_3 \end{bmatrix} \begin{bmatrix} 1 & 1 & 1 \\ 1 & 1 & 0 \\ 1 & 0 & 1 \\ 0 & 1 & 1 \end{bmatrix}$$

Therefore,

$$r_0 = (m_0 \times 1) \oplus (m_1 \times 1) \oplus (m_2 \times 1) \oplus (m_3 \times 0)$$
$$r_0 = m_0 \oplus m_1 \oplus m_2$$
$$r_1 = (m_0 \times 1) \oplus (m_1 \times 1) \oplus (m_2 \times 0) \oplus (m_3 \times 1)$$
$$r_1 = m_0 \oplus m_1 \oplus m_3$$
$$r_2 = (m_0 \times 1) \oplus (m_1 \times 0) \oplus (m_2 \times 1) \oplus (m_3 \times 1)$$
$$r_2 = m_0 \oplus m_2 \oplus m_3$$

- Finding the redundant or parity bits; when the message bit is $[m_0 \quad m_1 \quad m_2 \quad m_3] = [0 \quad 0 \quad 0 \quad 0]$

$$r_0 = m_0 \oplus m_1 \oplus m_2 = 0 \oplus 0 \oplus 0 = 0$$
$$r_1 = m_0 \oplus m_1 \oplus m_3 = 0 \oplus 0 \oplus 0 = 0$$
$$r_2 = m_0 \oplus m_2 \oplus m_3 = 0 \oplus 0 \oplus 0 = 0$$

Step 3:

Therefore, the final codeword =

m_0 m_1 m_2 m_3	r_0 r_1 r_2
0 0 0 0	0 0 0
Message	Parity bits

- Finding the redundant or parity bits, when the message bit is $[m_0 \quad m_1 \quad m_2 \quad m_3] = [0 \quad 0 \quad 0 \quad 1]$

$$r_0 = m_0 \oplus m_1 \oplus m_2 = 0 \oplus 0 \oplus 0 = 0$$
$$r_1 = m_0 \oplus m_1 \oplus m_3 = 0 \oplus 0 \oplus 1 = 1$$
$$r_2 = m_0 \oplus m_2 \oplus m_3 = 0 \oplus 0 \oplus 1 = 1$$

Therefore, the final codeword =

m_0 m_1 m_2 m_3	r_0 r_1 r_2
0 0 0 0	0 1 1
Message	Parity bits

Similarly, we can find the codewords for all the possible combinations of message bits. The complete list of message bits and codewords are listed in Table 7.1.

Table 7.1 Code words for (6,3) linear block code

Message bits				Parity bits			Codeword						
$[m_0$	m_1	m_2	$m_3]$	$[r_0$	r_1	$r_2]$	$[m_0$	m_1	m_2	m_3	r_0	r_1	$r_2]$
0	0	0	0	0	0	0	0	0	0	0	0	0	0
0	0	0	1	0	1	1	0	0	0	1	0	1	1
0	0	1	0	1	0	1	0	0	1	0	1	0	1
0	0	1	1	1	1	0	0	0	1	1	1	1	0
0	1	0	0	1	1	0	0	1	0	0	1	1	0
0	1	0	1	1	0	1	0	1	0	1	1	0	1
0	1	1	0	0	1	1	0	1	1	0	0	1	1
0	1	1	1	0	0	0	0	1	1	1	0	0	0
1	0	0	0	1	1	1	1	0	0	0	1	1	1
1	0	0	1	1	0	0	1	0	0	1	1	0	0
1	0	1	0	0	1	0	1	0	1	0	0	1	0
1	0	1	1	0	0	1	1	0	1	1	0	0	1
1	1	0	0	0	0	1	1	1	0	0	0	0	1

(Continued)

Table 7.1   Continued

Message bits				Parity bits			Codeword						
$[m_0$	m_1	m_2	$m_3]$	$[r_0$	r_1	$r_2]$	$[m_0$	m_1	m_2	m_3	r_0	r_1	$r_2]$
1	1	0	1	0	1	0	1	1	0	1	0	1	0
1	1	1	0	1	0	0	1	1	1	0	1	0	0
1	1	1	1	1	1	1	1	1	1	1	1	1	1

2. For a (6, 3) linear block code, whose generator matrix is given below. Realize an encoder for this code.

$$G = \begin{bmatrix} 1 & 0 & 0 & \vdots & 1 & 0 & 1 \\ 0 & 1 & 0 & \vdots & 0 & 1 & 1 \\ 0 & 0 & 1 & \vdots & 1 & 1 & 0 \end{bmatrix}$$

Solution:

The general format of linear block code is (n, k). It is compared with given problems (6, 3) linear block code. It gives, $n = 6$; and $k = 3$. It represents the total number of bits in message bits as 3 (i.e. $k = 3$) and the total number of bits in the codeword is 6 (i.e. $n = 6$). Hence, $(n - k)$ is the parity bits (i.e. $6 - 3 = 3$ bits are redundant bits).

Step 1:

Separate the coefficient matrix P from the given generator matrix G.
It is known that

$$G = [I_k \vdots P].$$

Comparing the above expression with given generator matrix G, we find

Identity matrix $I_k = \begin{bmatrix} 1 & 0 & 0 \\ 0 & 1 & 0 \\ 0 & 0 & 1 \end{bmatrix}$ and coefficient matrix $P = \begin{bmatrix} 1 & 0 & 1 \\ 0 & 1 & 1 \\ 1 & 1 & 0 \end{bmatrix}$

Step 2: Finding parity or redundant bits

These can be calculated using the following formulae.

$$R = M \bullet P \tag{7.14}$$

Here, R denotes redundant bits, M represents message bits, and P represents parity bits.

Here, the size of redundant and message bits are 3-bit long. Therefore, Equation (7.14) can be expressed as

$$[r_0 \quad r_1 \quad r_2] = [m_0 \quad m_1 \quad m_2] \begin{bmatrix} 1 & 0 & 1 \\ 0 & 1 & 1 \\ 1 & 1 & 0 \end{bmatrix}$$

Therefore,

$$r_0 = (m_0 \times 1) \oplus (m_1 \times 0) \oplus (m_2 \times 1)$$
$$r_0 = m_0 \oplus m_2$$
$$r_1 = (m_0 \times 0) \oplus (m_1 \times 1) \oplus (m_2 \times 1)$$
$$r_1 = m_1 \oplus m_2$$
$$r_2 = (m_0 \times 1) \oplus (m_1 \times 1) \oplus (m_2 \times 0)$$
$$r_2 = m_0 \oplus m_1$$

- Finding the redundant bits, when message bit is
  $[m_0 \quad m_1 \quad m_2] = [0 \quad 0 \quad 0]$

$$r_0 = m_0 \oplus m_2 = 0 \oplus 0 = 0$$
$$r_1 = m_1 \oplus m_2 = 0 \oplus 0 = 0$$
$$r_2 = m_0 \oplus m_1 = 0 \oplus 0 = 0$$

Therefore, the final codeword =

Message	Parity bits
0 0 0	0 0 0
m_0 m_1 m_2	r_0 r_1 r_2

We can obtain codewords for all the possible combination of message bits in a similar manner. Complete list of message bits and its codeword is listed in Table 7.2.

Table 7.2 Code words for (6,3) block codes

Message bits			Parity bits			Codeword					
$[m_0$	m_1	$m_2]$	$[r_0$	r_1	$r_2]$	$[m_0$	m_1	m_2	r_0	r_1	$r_2]$
0	0	0	0	0	0	0	0	0	0	0	0
0	0	1	1	1	0	0	0	1	1	1	0
0	1	0	0	1	1	0	1	0	0	1	1
0	1	1	1	0	1	0	1	1	1	0	1
1	0	0	1	0	1	1	0	0	1	0	1
1	0	1	0	1	1	1	0	1	0	1	1
1	1	0	1	1	0	1	1	0	1	1	0
1	1	1	0	0	0	1	1	1	0	0	0

The encoder for (6,3) block code is shown in Figure 7.3.

Encoder diagram:

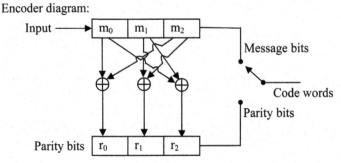

Figure 7.3   Encoder for (6,3) linear block code.

7.2.4 Decoding of Linear Block Codes (Receiver Side)

The steps involved in the decoding of linear block codes at receiver side are listed below:

1. Obtain the parity check matrix H:
   H is given by,
   $$H = \left[P^T | I_{(n-k)} \right]$$
   where P^T is the transpose of parity coefficient matrix
   $I_{(n-k)}$ is identity matrix
2. Find the transpose of parity check matrix H:
   i.e. $H^T = \begin{bmatrix} P \\ I_{n-k} \end{bmatrix}$
3. Multiply the received codewords Y with H^T
   - When the result of $Y \bullet H^T = (0, 0, 0, \ldots 0)$, there is no error.
   - When $Y \bullet H^T \neq (0, 0, 0, \ldots 0)$, the received codewords are affected by noise.

Solved problems on decoding of linear block codes:

1. For a $(6, 3)$ linear block code, coefficient matrix $[P]$ is as follows:
   $$[P] = \begin{bmatrix} 0 & 1 & 1 \\ 1 & 0 & 1 \\ 1 & 1 & 0 \end{bmatrix}$$. The received codewords at the receiver are
   (i) [0 0 1 1 1 0] and (ii) [1 1 1 0 1 1].
   Check whether they are correct or contain errors.
   Solution:
   Step 1: Find parity check matrix H.

where $H = \begin{bmatrix} P^T \mid I_{(n-k)} \end{bmatrix}$

$$
\left[
\begin{array}{ccc:ccc}
0 & 1 & 1 & 1 & 0 & 0 \\
1 & 0 & 1 & 0 & 1 & 0 \\
1 & 1 & 0 & 0 & 0 & 1 \\
\underbrace{}_{P^T} & & & \underbrace{}_{I_{(n-k)}} & &
\end{array}
\right]
$$

Step 2: Find H^T

$$
H^T = \begin{bmatrix} P \\ \hline I_{n-k} \end{bmatrix}
$$

$$
H^T = \left[
\begin{array}{ccc}
0 & 1 & 1 \\
1 & 0 & 1 \\
1 & 1 & 0 \\
\hline
1 & 0 & 0 \\
0 & 1 & 0 \\
0 & 0 & 1
\end{array}
\right]
$$

Step 3: Multiply the received codeword $[Y]$ by H^T
In the given problem, the received codewords are
$[Y_1] = [0\ 0\ 1\ 1\ 1\ 0]$ and $[Y_2] = [1\ 1\ 1\ 0\ 1\ 1]$

- $Y_1 H^T = [0\ 0\ 1\ 1\ 1\ 0]
\begin{bmatrix}
0 & 1 & 1 \\
1 & 0 & 1 \\
1 & 1 & 0 \\
1 & 0 & 0 \\
0 & 1 & 0 \\
0 & 0 & 1
\end{bmatrix}$

$$
= \begin{bmatrix}
(0 \times 0) \oplus (0 \times 1) \oplus (1 \times 1) \oplus (1 \times 1) \oplus (1 \times 0) \oplus (0 \times 0), \\
(0 \times 1) \oplus (0 \times 0) \oplus (1 \times 1) \oplus (1 \times 0) \oplus (1 \times 1) \oplus (0 \times 0), \\
(0 \times 1) \oplus (0 \times 1) \oplus (1 \times 0) \oplus (1 \times 0) \oplus (1 \times 0) \oplus (0 \times 0)
\end{bmatrix}
$$

$= [0 \oplus 0 \oplus 1 \oplus 1 \oplus 0 \oplus 0, 0 \oplus 0 \oplus 1 \oplus 0 \oplus 1 \oplus 0, 0 \oplus 0 \oplus 0 \oplus 0 \oplus 0 \oplus 0]$
$= [0,\ 0,\ 0]$

Here, $Y_1 H^T = [0,\ 0,\ 0]$
Therefore, there is no error in the received codeword.

$$\bullet \ Y_2 H^T = \begin{bmatrix} 1 & 1 & 1 & 0 & 1 & 1 \end{bmatrix} \begin{bmatrix} 0 & 1 & 1 \\ 1 & 0 & 1 \\ 1 & 1 & 0 \\ 1 & 0 & 0 \\ 0 & 1 & 0 \\ 0 & 0 & 1 \end{bmatrix}$$

$$= \begin{bmatrix} (1 \times 0) \oplus (1 \times 1) \oplus (1 \times 1) \oplus (0 \times 1) \oplus (1 \times 0) \oplus (1 \times 0), \\ (1 \times 1) \oplus (1 \times 0) \oplus (1 \times 1) \oplus (0 \times 0) \oplus (1 \times 1) \oplus (1 \times 0), \\ (1 \times 1) \oplus (1 \times 1) \oplus (1 \times 0) \oplus (0 \times 0) \oplus (1 \times 0) \oplus (1 \times 1) \end{bmatrix}$$

$$= [0 \oplus 1 \oplus 1 \oplus 0 \oplus 0 \oplus 0, 1 \oplus 0 \oplus 1 \oplus 0 \oplus 1 \oplus 0, \ 1 \oplus 1 \oplus 0 \oplus 0 \oplus 0 \oplus 1]$$
$$= [0, \ 1, \ 1]$$

Here, $Y_2 H^T \neq [0, \ 0, \ 0]$

Therefore, the received codeword is affected by noise.

7.2.5 Error Correction (Syndrome Decoding)

Error correction capability of linear block code is obtained by *Syndrome decoding*.

Syndrome is given by $S = YH^T$

Steps involved in error correction:

1. For the given received codeword, obtain the syndrome.

$$S = YH^T$$

2. Now compare the syndrome S with each row of H^T
3. When there is a match, identify the row number. This row number indicates the location of error or bit position of error
4. Invert the corresponding bit.

7.2.6 Solved Problems on Syndrome Decoding

Solved Problems on Error correction (Syndrome decoding):

1. For a $(6, 3)$ linear block code, the received codeword is $[1 \ 1 \ 1 \ 0 \ 1 \ 1]$. Check whether the received codeword is affected by noise. If it is

affected by noise, identify the bit position and correct the error. The coefficient matrix is given below:

$$[P] = \begin{bmatrix} 0 & 1 & 1 \\ 1 & 0 & 1 \\ 1 & 1 & 0 \end{bmatrix}$$

Solution:
Step 1: Find the syndrome

$$S = YH^T \quad Here, Y = [1\ 1\ 1\ 0\ 1\ 1]$$

Step 2: Find H^T

$$H^T = \left[\frac{P}{I_{n-k}} \right]$$

$$H^T = \begin{bmatrix} 0 & 1 & 1 \\ 1 & 0 & 1 \\ 1 & 1 & 0 \\ \hline 1 & 0 & 0 \\ 0 & 1 & 0 \\ 0 & 0 & 1 \end{bmatrix}$$

$$S = [1\ 1\ 1\ 0\ 1\ 1] \begin{bmatrix} 0 & 1 & 1 \\ 1 & 0 & 1 \\ 1 & 1 & 0 \\ \hline 1 & 0 & 0 \\ 0 & 1 & 0 \\ 0 & 0 & 1 \end{bmatrix}$$

$$= \begin{bmatrix} (1 \times 0) \oplus (1 \times 1) \oplus (1 \times 1) \oplus (0 \times 1) \oplus (1 \times 0) \oplus (1 \times 0), \\ (1 \times 1) \oplus (1 \times 0) \oplus (1 \times 1) \oplus (0 \times 0) \oplus (1 \times 1) \oplus (1 \times 0), \\ (1 \times 1) \oplus (1 \times 1) \oplus (1 \times 0) \oplus (0 \times 0) \oplus (1 \times 0) \oplus (1 \times 1) \end{bmatrix}$$

$$= [0 \oplus 1 \oplus 1 \oplus 0 \oplus 0 \oplus 0, 1 \oplus 0 \oplus 1 \oplus 0 \oplus 1 \oplus 0, 1 \oplus 1 \oplus 0 \oplus 0 \oplus 0 \oplus 1]$$
$$= [0,\ 1,\ 1]$$
$$S = [0,\ 1,\ 1]$$

Step 3: Compute the value of S with H^T

$$S = [0,\ 1,\ 1] \quad H^T = \begin{bmatrix} 0 & 1 & 1 \\ 1 & 0 & 1 \\ 1 & 1 & 0 \\ \hline 1 & 0 & 0 \\ 0 & 1 & 0 \\ 0 & 0 & 1 \end{bmatrix}$$

Now, the computed syndrome S matches with the first row of H^T. It informs that, first bit in the received codeword is affected by noise.

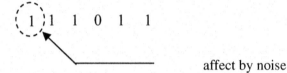

affect by noise

Step 4: Error correction
Invert the affected bit position. Hence, the corrected codeword is, 0 1 1 0 1 1

7.3 Convolutional Codes

It is one of the error-correcting codes and can be defined by the following parameters: (n, k, K), where n represents encoder output, k is the number of input bits, and K is the *constraint length*
Constraint length:
    It refers to the output of encoder and depends on $(k - 1)$ previous inputs. In other words, it refers to the number of flip-flops in the shift register.
We can decode the convolutional encoder at the receiver side using the *Viterbi decoding algorithm*.
The operation of convolutional encoder can be done using three methods: They are: (1) State diagram, (2) Trellis diagram, and (3) Code tree approach.

7.3.1 Solved Problems

1. Convolutional encoder is shown in the figure below. Find the output of the encoder when the input is 0 1 0 1 1 0 0 0 1 1 1. Also discuss the operation of the convolutional encoder with the help of a state diagram, Trellis diagram and the code tree.

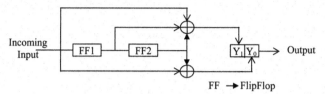

Figure 7.4 Convolutional encoder.

Solution:

The output of the encoder is

$$Y_1 = \text{input} \oplus FF_1 \oplus FF_2$$

$$Y_0 = \text{input} \oplus FF_2$$

We need to find the output of the encoder when the input to the encoder is, 0 1 0 1 1 0 0 0 1 1 1

Let us assume that initial states of flip-flop FF_1 and FF_2 are 0 and 0 as shown in Figure 7.5.

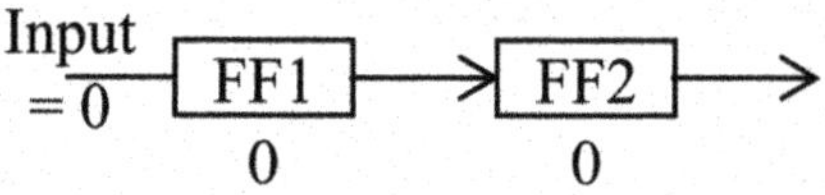

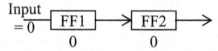

Figure 7.5 Initial state of flip flop FF1 and FF2.

State 1: Applying the first input

When we apply first input 0 to the encoder, i.e. input $= 0$ is shown in Figure 7.6.

Figure 7.6 Applying first input.

During this stage, the output of the encoder is

$$Y_1 = \text{input} \oplus FF_1 \oplus FF_2 = 0 \oplus 0 \oplus 0 = 0$$
$$Y_0 = \text{input} \oplus FF_2 = 0 \oplus 0 = 0$$

Therefore, the output of the encoder is 0 0, when we apply input $= 0$.

Stage 2: Applying the second input

The current state of the encoder is shown in Figure 7.7.

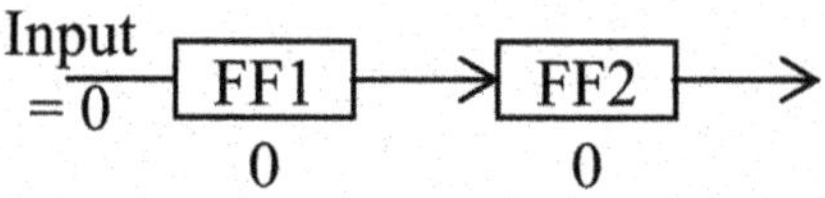

Figure 7.7 Applying second input to convolution encoder.

When we apply the second input as 1, i.e. input = 1; the current state becomes modified into the following manner shown in Figure 7.8.

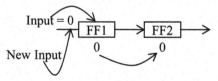

Figure 7.8 Modification of current state of FF1 and FF2.

When we are ready to apply the second input, the current input = 0 is moved to FF_1 and content of FF_1 is moved to FF_2. This operation is given in a pictorial way by the following Figure 7.9.

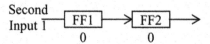

Figure 7.9 Modification of current state of FF1 and FF2; When apply second input.

Now the output of the encoder is

$$Y_1 = \text{input} \oplus FF_1 \oplus FF_2 = 1 \oplus 0 \oplus 0 = 1$$
$$Y_0 = \text{input} \oplus FF_2 = 1 \oplus 0 = 1$$

Stage 3: Applying the third input
When we apply the third input as 0, the current bit in the input side is moved to FF_1, and content of FF_1 is moved to FF_2 which is shown in Figure 7.10.

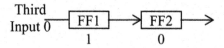

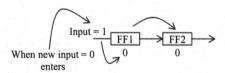

Figure 7.10 Modification of current state of FF1 and FF2; When apply third input.

Now the output of the encoder is

$$Y_1 = \text{input} \oplus FF_1 \oplus FF_2 = 0 \oplus 1 \oplus 0 = 1$$
$$Y_0 = \text{input} \oplus FF_2 = 0 \oplus 0 = 0$$

Stage 4: Applying the fourth input
When we apply the fourth input as 1, the current bit in the input side is moved to FF_1 and content of FF_1 is moved to FF_2 which is shown in Figure 7.11.

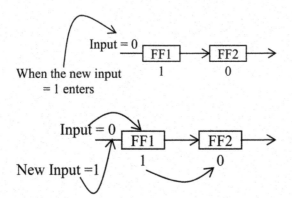

Figure 7.11 Modification of current state of FF1 and FF2; When apply fourth input.

Therefore, the states of the Flip–flops are mentioned in Figure 7.12.

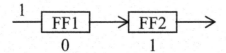

Figure 7.12 Condition of flip flop.

Now the output of the encoder is

$$Y_1 = \text{input} \oplus FF_1 \oplus FF_2 = 1 \oplus 0 \oplus 1 = 0$$
$$Y_0 = \text{input} \oplus FF_2 = 1 \oplus 1 = 0$$

We can obtain all the outputs of the encoder in a similar manner. For clear understanding, the remaining outputs of the encoder are given in the state Table 7.3.
State table of encoder:

Table 7.3 Summary of encoder operation

Input	Current state $FF_1\ FF_2$	Next state $FF_1\ FF_2$	Output $Y_1\ Y_0$
Initial condition	0 0	– –	– –
0	0 0	0 0	0 0
1	0 0	1 0	1 1
0	1 0	0 1	1 0

(Continued)

Table 7.3 Continued

Input	Current state $FF_1 FF_2$	Next state $FF_1 FF_2$	Output $Y_1 Y_0$
1	0 1	1 0	0 0
1	1 0	1 1	0 1
0	1 1	0 1	0 1
0	0 1	0 0	1 1
0	0 0	1 0	1 1
1	0 0	1 0	1 1
1	1 0	1 1	0 1
1	1 1	1 1	1 0

Therefore,

The incoming input

0   1   0   1   1   0   0   0..............

The encoder output

00   11   10   00   01   01   11   00...........

State diagram:

The operation of the convolutional encoder is better explained with state diagram. Given that the convolutional encoder consists of two flip-flops. Therefore, four different states are possible. They are state $a = 00$, state $b = 01$, state $c = 10$, and state $d = 11$

Table 7.4 gives the next state of the flip-flop.

Table 7.4 State table

Current State	Next State When input = 0	When input = 1
$a = 00$	00	10
$b = 01$	00	10
$c = 10$	01	11
$d = 11$	01	11

State diagram:

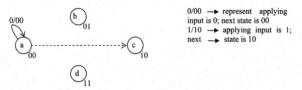

0/00 → represent applying input is 0; next state is 00
1/10 → applying input is 1; next → state is 10

Figure 7.13 State diagram of a encoder.

The Figure 7.13 gives the state diagram, when the current state = 00 (a), incoming input may be 0 or 1. When it is 0, the next state is 00 (same state), hence it is marked as a solid line. When the incoming input = 1, the next state is 10 and is marked as a dotted line.

We can draw the state transitions of the states also in a similar manner. The complete state diagram is shown in the Figure 7.14.

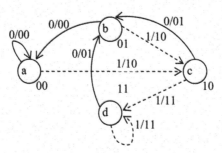

Figure 7.14 Complete state diagram of a encoder.

Operation of convolutional encoder with the aid of Trellis diagram:

The information contained in the state diagram can be conveyed graphically. Graphical format of representation is *"Trellis diagram"*.

In the Figure 7.15, left-side nodes are the possible current states and the right-side nodes are the possible next states.

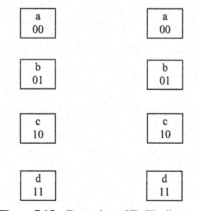

Figure 7.15 Formation of Trellis diagram.

Case 1:

For example, let us assume the current state of the encoder as state $a = 00$, the possible next state is either 00 or 10 and based on the incoming input (i.e. when the input = 0, the next state = 00, when input = 1, the next state = 10). These information are drawn in Trellis diagram by the following Figure 7.16:

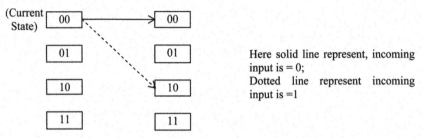

Figure 7.16　First stage of Trellis diagram.

Case 2:

Let us assume, the current state of the encoder as 10 (state *c*). What would be the Trellis diagram?

Answer: As we know, the next state of the encoder is 01, when the incoming input = 0; otherwise, the next state of the encoder is 11, when we apply input = 11.

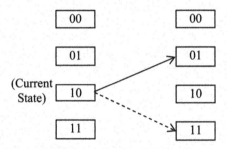

Figure 7.17　Second stage of Trellis diagram.

In a similar manner, we can draw all the possible combinations. The complete Trellis diagram is shown in the Figure 7.18.

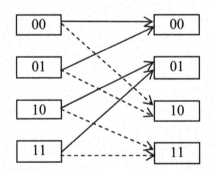

Figure 7.18　Trellis for the convolutional encoder.

2. *Draw the Trellis diagram for the convolutional encoder shown in Figure 7.4, when the input to the encoder is 0 1 1.*

Solution:

Stage 1:

Assume that the initial state of the encoder is 00 (i.e. state *a*). During this stage apply the first input 0. Trellis diagram for this stage is given in the Figure 7.19.

Figure 7.19 First stage of Trellis diagram.

For a complete Trellis diagram, we must draw all the possible combinations, apart from the current state and the next state. It is shown in the Figure 7.20.

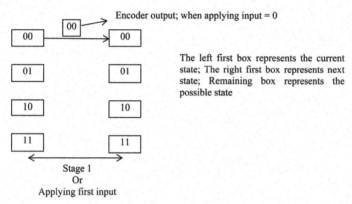

Figure 7.20 Trellis diagram; when applying first input.

Stage 2: Applying the second input = 1

At stage 2, the current state = 00, during this stage when we apply the second input = 1, the next state becomes 10, this is given in the Figure 7.21.

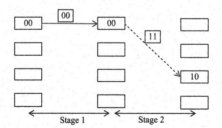

Figure 7.21 Trellis diagram; when applying second input.

Stage 3: Applying the third input = 1

At stage 3, the current state = 10, during this stage, when we apply the third input = 1, the next state becomes 11. This is given in the Figure 7.22.

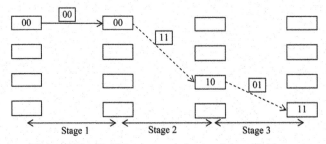

Figure 7.22   Trellis structure; when applying third input.

Block color box represents current and the next state, and blue box represents the possible state combinations.

Code tree:

This is another approach to explain the operation of a convolutional encoder.

1. Draw the code tree for the convolutional encoder shown in the Figure 7.4. When the current state of the encoder is 00.

Solution:

We start with the current state 00, the next state becomes 00, when the input = 0, otherwise the next state = 10. State table for the given convolutional encoder is shown in Table 7.5.

State table:

Table 7.5   State table

Current State	Next State When input = 0	When input = 1
$a = 00$	00	10
$b = 01$	00	10
$c = 10$	01	11
$d = 11$	01	11

Code tree at stage 1 is shown in Figure 7.23.

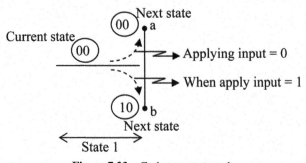

Figure 7.23   Code tree at stage 1.

This code tree describes the operation of the encoder. When the incoming input = 0, the corresponding transitions are marked in upward direction, otherwise (input = 1) the direction is downward.

Stage 2:

Now there are two possible states 00 and 10. At 00, the next state = 00 and 10 based on the incoming input 0 or 1. Similarly, at point *b*, the next state = 01 or 11 based on the incoming input 0 or 1. This explanation is given in code tree Figure 7.24.

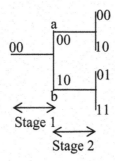

Figure 7.24 Code tree at stage 2.

In a similar manner, we can draw for other stages also and it is shown in Figure 7.25.

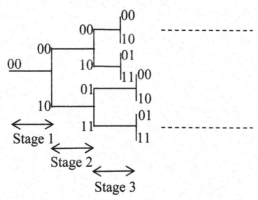

Figure 7.25 Complete code tree.

Summary: In the code tree, upward direction implies application of input = 0. Downward directions are for applying input = 1.

7.3.2 Viterbi Decoding Algorithm

It is the minimum length decoder or maximum likelihood decoding. It is the optimum decoding technique for a memoryless channel.

Basic terminology in Viterbi decoding:

Before studying Viterbi decoding algorithm, we must learn some basic terms.

What is Hamming distance and how is it calculated?

Hamming distance refers to the number of bit position difference between two codewords. For example, let us take the codewords 1 1 0 0 and 1 0 0 1. Here, Hamming distance is 2.

Since 1 1 0 0

  1 0 0 1.

The number of bit position difference is two. Therefore, the Hamming distance is two.

- Find the Hamming distance between 1 0 and 0 1.

Answer: 1 0

  0 1

The number of bit position difference is two. Therefore, the Hamming distance = 2.

- Find the Hamming distance between 0 0 and 1 0.

Answer: 0 0

  1 0

The number of bit position difference is one. Therefore, the Hamming distance = 1.

Another important point is that we need to calculate "*path metric*" and "*branch metric*" in Viterbi decoding algorithm.

Path metric calculation:

Path metric:

It is defined as the Hamming distance between the received signal (Y) and the decoded signal at a specific node. It is represented by a number covered by brackets.

- Find the path metric for the given Trellis diagram shown in Figure 7.26.

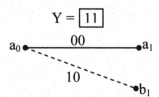

Figure 7.26 Trellis diagram.

Answer:

Given Y = 11 is a received signal.

Path metric between nodes a_0 and a_1:

Received signal = 1 1

Decoded signal at node a_0 to a_1 = $\underline{00}$

The number of bit position difference is two, i.e. Hamming distance = 2. Path metric = 2.

Path metric between nodes a_0 and b_1:

Received signal = 1 1

Decoded signal at node a_0 to b_1 = 1 $\underline{0}$

The number of bit position difference is one. Path metric = 1.

In pictorial way, it can represented by the Figure 7.27

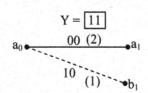

Figure 7.27   Mentioning Path metric in Trellis diagram.

Find the path metric for the Figure 7.28

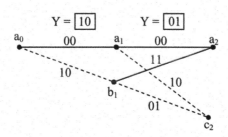

Figure 7.28   Trellis diagram.

Answer:

Stage 1: *Path metric between nodes a_0 and a_1:*

Received signal = $\underline{1}$ 0

Decoded signal at node a_0 to a_1 = $\underline{0}$ 0

The number of bit position difference is one. Path metric = (1).

Path metric between nodes a_0 and b_1:

Received signal = 1 0

Decoded signal at node a_0 to b_1 = 1 0

The number of bit position difference is zero. Path metric = (0).

Stage 2:
Path metric between nodes a_1 and a_2:
Received signal = $\underline{0}$ 1
Decoded signal at node a_1 to a_2 = $\underline{0}$ 0
The number of bit position difference is one. Path metric = (1).
Path metric between nodes b_1 and c_2:
Received signal = 0 1
Decoded signal at node b_1 to c_2 = 0 0
The number of bit position difference is zero. Path metric = (0).
Path metric between nodes b_1 and a_2:
Received signal = $\underline{0}$ 1
Decoded signal at node b_1 to a_2 = $\underline{1}$ 1
The number of bit position difference is one. Path metric = (1).
Path metric between nodes a_1 and c_2:
Received signal = $\underline{01}$
Decoded signal at node b_1 to a_2 = $\underline{10}$
The number of bit position difference is two. Path metric = (2). The complete
path metric mentioning is shown in Figure 7.29.

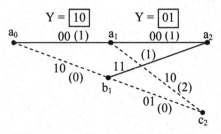

Figure 7.29   Path metric mentioning in Trellis diagram.

Branch metric calculation:

- Find the branch metric of the given Trellis diagram shown in Figure 7.30.

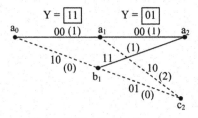

Figure 7.30   Trellis diagram.

Branch metric calculation:

Branch metric at node a_1 = Path metric at node a_0 + Path metric between nodes a_0 and a_1 = $(0) + (1)$ = ①

Branch metric at node a_2 = Branch metric at node a_1 + Path metric between nodes a_1 and a = ①(1) = ②

Branch metric at node b_1 = Path metric at node a_0 + Path metric between nodes a_0 and b_1 = $(0) + (0)$ = ⓪

Branch metric at node a_2 from b_1 = Branch metric at node b_1 + Path metric between nodes b_1 and a_2 ⓪(1) = ①

Branch metric at node c_2 from b_1 = Branch metric at node b_1 + Path metric between nodes b_1 and c_2 = ⓪(0) = ⓪ ①

Branch metric at node c_2 from a_1 = Branch metric at node a_1 + Path metric between nodes a_1 and c_2 ①(2) = ③

This can be represented pictorially by the Figure 7.31:

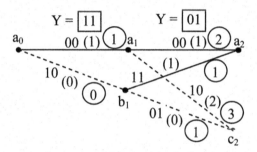

Figure 7.31 Path metric and Branch metric for Trellis diagram.

Surviving path:

At the last stage of Viterbi decoding in Trellis diagram, each node ends up with two paths of which only one has a minimum path metric. This path is referred to as "*Surviving path*". For example, let us take the Trellis diagram shown in Figure 7.32.

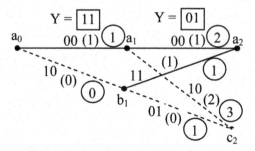

Figure 7.32 Trellis diagram.

In this figure, at final stage, node a_2 and c_2 end up with two paths.
i.e. Node a_2 paths are (i) (a_0 *to* a_1) and (a_1 *to* a_2) (ii) (a_0 *to* b_1) and (b_1 *to* c_2)
The total path metric for path 1 is
(a_0 to a_1) and (a_1 to a_2) = ① ② ③
Total path metric for path 2 is
(a_0 to b_1) and (b_1 to c_2) = ⓪①①
Hence, path 2 here is a surviving path.
A similar calculation can be done at node c_2; node c_2 ends up with following
paths:
(i) (a_0 *to* b_1) and (b_1 *to* c_2) and (ii) (a_0 *to* a_1) and (a_1 *to* c_2)
Path metric for path 1 is = ⓪ = ⓪ ⓪
Path metric for path 2 is = ① = ③ ④
Hence, path 1 here is the surviving path.

7.3.3 Solved Problems

(i) Draw the State diagram and the Trellis diagram for a convolutional
   encoder Figure 7.33.
(ii) Find the output sequence of an encoder when the inputs are 0 1 0 1 1 0.
(iii) Use Viterbi algorithm to decode the sequence 00 11 01 01

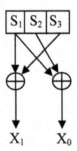

Figure 7.33   Convolutional encoder.

Solution:
In the given convolutional encoder circuit, the number of inputs $(k) = 1$ and
number of outputs $(n) = 2$
Also, the output of the encoder depends on previous two inputs. Therefore,
constraint length $(K) = 3$
We can represent this mathematically using the following notations:
$(n, k, K) = (2, 1, 3)$

The output of the circuit is

$$X_1 = \text{input} \oplus s_3$$

$$X_0 = \text{input} \oplus s_2$$

The output of the circuit depends on previous two inputs at any point of time. Therefore, four different states are possible. They are a $= 00$; b $= 01$; c $= 10$; d $= 11$. Summary of encoder operation is shown in Table 7.6. State table:

Table 7.6 Summary of encoder operation

Input S_1	Current State S_2S_3	Next State S_2S_3	Output X_1X_0
Initial condition	0 0	– –	– –
0	0 0	0 0	0 0
1	0 0	1 0	1 1
0	0 1	0 0	1 0
1	0 1	1 0	0 1
0	1 0	0 1	0 1
1	1 0	1 1	1 0
0	1 1	0 1	1 1
1	1 1	1 1	0 0

The present state and the next state table is shown in Table 7.7 and Figure 7.34 indicates the state diagram of a given convolutional encoder:

Table 7.7 State table

	Next State	
Current State	When input = 0	When input = 1
$a = 00$	00	10
$b = 01$	00	10
$c = 10$	01	11
$d = 11$	01	11

State diagram:

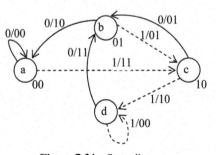

Figure 7.34 State diagram.

In the state diagram, the solid line represents input = 0; dotted line represents input = 1.

$0/10$ informs that numerator 0 indicates input = 0 and denominator 10 represents the encoder output.

Trellis diagram:

The Figure 7.35 is the Trellis diagram for a given convolutional encoder.

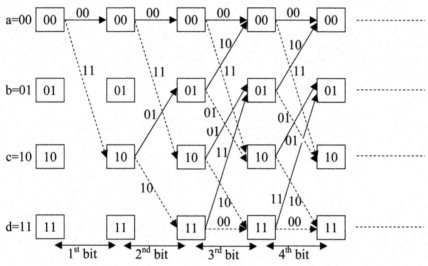

Figure 7.35 Trellis diagram.

(ii) *Encoder output when the inputs are 0 1 0 1 1 0.*

Based on the knowledge of constructing state table which is shown in Table 7.8, we can draw the State diagram for the given input sequence.

Table 7.8 Summary of encoder operation

Input S_1	Current State S_2S_3	Next State S_2S_3	Output X_1X_0
Initial condition	0 0	- -	- -
0	0 0	0 0	0 0
1	0 0	1 0	1 1
0	1 0	0 1	0 1
1	0 1	1 0	0 1
1	1 0	1 1	1 0
0	1 1	0 1	1 1

Input sequence: 0   1   0   1   1   0

Output encoded sequence: 00   11   01   01   10   11

In another way, we can determine the output sequence based on Trellis diagram:

- When the input = 0;

The Trellis diagram for a given convolutional encoder is shown in Figure 7.36 when the input = 0

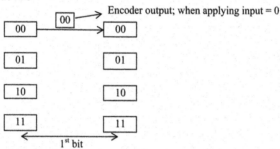

Figure 7.36 Trellis diagram; when apply first input.

Therefore, when the input is zero, the encoded output is 0 0.

- When the input = 1:

The second stage of Trellis diagram is shown in Figure 7.37.

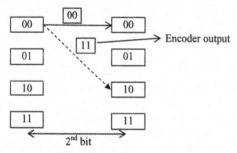

Figure 7.37 Second stage of Trellis diagram.

When the input = 1, encoded output = 1 1
When the input = 0, the trellis diagram is shown in Figure 7.38.

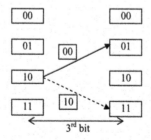

Figure 7.38 Third stage of Trellis diagram.

The same operation can be repeated for any input applied to the convolutional encoder.

Viterbi decoding:

In this Figure 7.39, minimum path metric path is $(a_0 - a_1)$ to $(a_1 - c_2)$ to $(c_2 - b_3)$ to $(b_3 - c_4)$

Decoder output is,

00	11	01	01
0	1	0	1

$$(a_0 - a_1)(a_1 - c_2)(c_2 - b_3)(b_3 - c_4)$$

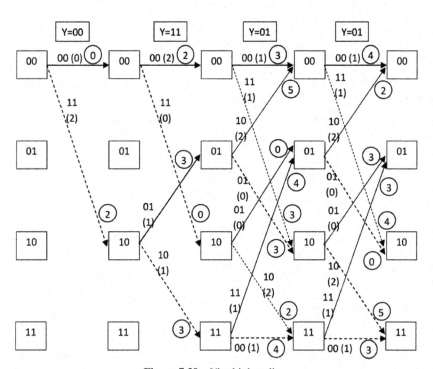

Figure 7.39   Viterbi decoding.

PART V

Multi-User Radio Communication

8

Wireless Communication

8.1 Introduction

Today mobile phones are used not only for making calls, but also for finding stock quotes, online games, real-time money transfer, etc. It implies data traffic today being more in the telecommunication industry compared with voice traffic. Mobile communications are divided into various generations for supporting data traffic and for better quality of service (QoS) to the user. They are,

1G System
- 1G system supports only analog voice services. It cannot provide data service to the user.
- It was introduced in 1980s and completed in early 1990s.

Examples
1. NMT-Nordic Mobile Telephone is used in Nordic countries, Switzerland.
2. AMPS-Advanced mobile phone system is used in North America and Australia.
3. TACS-Total access communication systems are used in United Kingdom.

Limitations
- Poor voice quality
- Poor battery life
- Large phone size
- No security
- Poor handoff reliability

2G Systems
- 2G networks use digital signals based on Global system for mobile (GSM).
- It was launched in Finland in 1991.
- It can support low-speed data service in the range of 64 Kbps.

Features

- It enables services such as text message, picture message, and MMS (multimedia message).
- It offers better quality and capacity.

Examples: GSM, CDMA, and D-AMPS

2.5 G systems

It is described as 2G cellular technology combined with GPRS.

Features

- Phone calls
- Send/receive email message
- Web browsing
- It offers data services in the range of 64–144 Kbps

Examples: GPRS, CDMA 2000 1x

3G Networks

- It was introduced in 2000.
- It offers data rate in the range of 384 Kbps to 2 Mbps

Features

- It provides faster communication
- Send/receive large email message
- More security
- Video conferencing

Examples: UMTS, EDGE, and HSPA

8.2 Advanced Mobile Phone System

- It is a first-generation analog mobile cell phone system standard developed by Bell laboratories and introduced in the year 1983.
- It occupies 40 MHz spectrum in the 800 MHz band allocated by Federal communication commission (FCC).
- In 1989, an additional 10 MHz spectrum was allocated. It is referred to as an extended AMPS system.
- It uses frequency modulation (FM) for voice modulation.
- Mobile to BS (base station) (reverse link) uses frequencies between 824 and 849 MHz.
- BS to mobile (forward link) uses frequencies between 869 and 894 MHz.
- Separation for forward and reverse channels is 45 MHz.

8.2.1 AMPS Architecture

Figure 8.1 is an illustration of the AMPS system.

- Each BS has a control channel transmitter (broadcasts on the forward control channel), a control receiver (that listens on the reverse control channel for any cellular phone trying to set up a control), and 8 or more FM duplex voice channels. BS supports as many as 57 voice channels.

- Forward Voice Channel (FVC) carries the portion of the tele-conversation originating from the landline tele-network caller and going to the mobile user.

- Reverse voice channels (RVC) carry portions of the tele-conversation originating from the mobile user and going to a landline tele-network caller.

- The number of control and voice channels used at a BS varies based on load of the system.

- Each BS continuously transmits digital FSK data on the Forward Control Channel at all times so that idle mobile users can lock onto the strongest FCC wherever they are.

- All users must be locked or "camped" onto an FCC in order to originate or receive calls.

- The BS reverse control channel (RCC) receiver constantly monitors transmissions from mobile users that are locked onto the matching FCC.

- There are 21 control channels, which are scanned for finding the best serving base station.

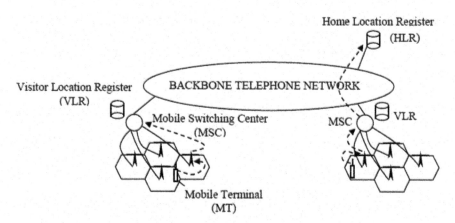

Figure 8.1 AMPS system architecture.

A Wired User calls a Mobile User

- The call arrives at MSC and a paging message is sent with the mobile's MIN (Mobile Identification Number) simultaneously on every BS forward control channel in the system.
- When the target caller receives the page, it responds with ACK transmission on the reverse control channel.
- MSC then directs the BS to assign a FVC and RVC pair to the mobile so that this call can take place on a dedicated voice channel.
- BS also assigns the mobile a Supervisory Audio Tone (SAT tone) and a Voice Mobile Attenuation Code (VMAC) as it moves the call to the voice channel.
- The mobile automatically changes its frequency to the assigned voice channel pair.

Mobile User places a Call

- It transmits a message on the RCC containing its MIN, Electronic Serial Number (ESN), Station Class Mark (SCM), and the destination telephone number.
- When received correctly by the BS, this information is sent to the MSC, which checks to ensure the mobile is properly registered, connects the mobile to the public switched telephone network (PSTN), assigns the call to a forward and reverse voice channel pair with a specific SAT and VMAC, and commences the conversation.
- Handoff decisions are made by the MSC when the signal strength on the reverse voice channel (RVC) of the serving BS drops below a preset threshold or when the SAT tone experiences a certain level of interference.
- MSC uses scanning receivers called "locate receivers" in nearby BSs to determine the signal level of a particular user, which needs a handoff.
- In doing so, the MSC is able to find the best neighboring BS, which can accept the handoff.
- Mobile notifies a serving MSC its presence and location by periodically keying up and transmitting its identity information, which allows the MSC to constantly update its customer list.
- A Registration command is sent in the overhead message of each control channel at 5–10 min intervals and includes a timer value which each mobile uses to determine the precise time at which it should respond to the serving BS with a registration transmission.
- Each mobile reports its MIN and ESN during a brief registration transmission to enable MSC validating and updating the customer list.

Summary of AMPS standard

Multiple access:	FDMA
Duplexing:	FDD
Channel Bandwidth:	30 kHz
Reverse channel frequency:	824–849 MHz
Forward channel frequency:	869–894 MHz
Voice modulation:	FM
Number of channels:	666–832
Coverage radius by 1 base station:	2–20 km.

8.3 Global System for Mobile

- The GSM is a set of ETSI standards specifying the infrastructure for a digital cellular service.
- This standard is used in approximately 85 countries in the world including such locations as Europe, Japan, Australia, and also in India.

History of GSM

- The first GSM network was launched in 1991 and most European GSM networks turned commercial in 1992.
- 1994: Data transmission launched. 1995: About 156 members from 86 countries form networks, 1996: 120 networks in 71 other countries launched, 1997: 200 GSM networks in 109 countries. So exponential raise in the application and deployment of GSM networks. Now, we have 44 million subscribers in counting.

8.3.1 GSM System Hierarchy

GSM networks are structured hierarchically. They consist of an administrative region, which is assigned to a mobile switching center (MSC). Each administrative region is made up of at least one location area (LA), which is also called a visited area and consisting of several cell groups, each assigned to a base station controller (BSC). Cells of one BSC may belong to different LAs. Figure 8.2 shows the GSM system hierarchy.

8.3.2 GSM System Architecture

The GSM network can be divided into four main parts.

1. The mobile station,
2. The base station subsystem,

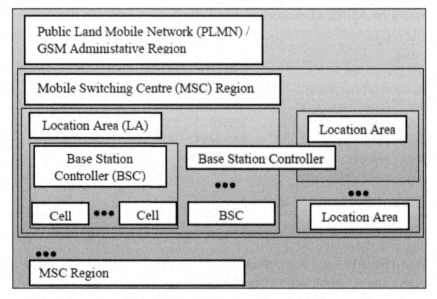

Figure 8.2   GSM system hierarchy.

3. The network and switching subsystem,
4. Operation and support subsystem.

The complete system architecture is shown in Figure 8.3.

1. Mobile station (MS)

A mobile station (MS) consists of two main elements,
   i) the equipment or terminal or mobile device and
   ii) the subscriber identity module (SIM).

i) The Terminal

- There are different types of terminals distinguished principally by their power and application:
- The *fixed* terminals are the ones installed in cars. Their maximum allowed output power is 20 W.
- The GSM portable terminals can also be installed in vehicles. Their maximum allowed output power is 8 W.
- The handheld terminals using these weights are decreasing drastically. The size and the weight of the handsets are going down every day. These terminals emit up to 2 W peek power and the maximum allowed power can go down to 0.8 W.

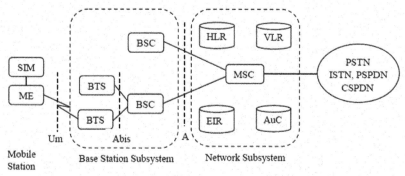

SIM   Subscriber Identity Module   BSC   Base Station Controller   MSC   Mobile Services Switching Center
ME   Mobile Equipment   HLR   Home Location Register   EIR   Equipment Identity Register
BTS   Base Transceiver Station   VLR   Visitor Location Register   AuC   Authentication Center

Figure 8.3   GSM system architecture.

ii) The SIM

- SIM is a smart card that identifies the terminal by inserting the SIM card into the terminal, the user can have access to all the subscribe services.
- The terminal cannot work without the SIM card which is protected by a 4-digit personalized identification number (PIN).
- Now, in order to identify the subscriber to the system, the SIM card contains some parameters of the users such as the International Mobile Subscriber Identity (IMSI) and another advantage of SIM is the mobility of the user.
- The SIM achieves separation of the user mobility from the equipment mobility. This enables international roaming independent of the mobile equipment and the network technology.
- All cryptographic algorithms for authentication and data encryption can be realized in the SIM and so the newer versions of the SIM can claim that your data will be more secure as the next-generation cryptographic algorithms will be embedded in the SIM.
- The SIM can also store short messages, charging information, list of important phone numbers for fast dialing, and several other things basically equipped with memory.

2) The Base station subsystem (BSS)

The base station subsystem (BSS) connects the mobile station (MS) and the network subsystem (NSS). It is in charge of the transmission and reception. The BSS can again be divided into two sub-parts,

i) The base transceiver station (BTS) or base station (BS), and

ii) Base station controller (BSC)

i) The Base transceiver station (BTS)

- The BTS corresponds to the transceivers and antennas used in each cell of the network. A BTS is usually placed in the center of the cell.
- Its transmitting power defines the size of the cell. Each BTS has between 1 and 16 transceivers depending upon the density of users in the cell.

ii) Base station controller (BSC)

- The BSC controls a group of BTS and manages their radio resources.
- A base station controller is principally in charge of handovers, frequency hopping, exchange functions, and control of the radio frequency power levels of the BTS.

3) The Network and Switching subsystem (NSS)

Its main role is to manage the communication between the mobile users and other users such as the mobile user, ISDN users, fixed telephone users, etc. It also includes databases needed in order to store information on the subscriber and to manage their mobility.

The different components of the NSS are:

i) The mobile services switching center (MSC)

- It is the central component of the NSS. The mobile switching center performs the switching functions for the network. It also provides connections to the other networks primarily the PSTN or the public switched telephone network.
- The MSC performs routing path search, signal routing, and service feature processing.

ii) Home and Visitor registers (HLR and VLR)

The GSM network has several databases. Two functional units are defined for the registration of subscribers and their current location. They are

- Home location registers (HLR) and
- Visitor location registers (VLR)

HLR

- There is a home location register (HLR) for every public land mobile network and one VLR in each of the MSC's. This arrangement depends on the number of subscribers processing and the storage capacity of the switches and the entire network structure.

- The HLR has entries for every subscriber and every mobile ISDN number, which has its home in the respective network. It stores permanent subscriber data and relevant temporary data or the subscribers permanently registered in the HLR.
- It also stores the current MS location. All administrative activities of the subscribers are performed in the database of HLR.

VLR

The VLR stores the data of all mobile stations, which are currently staying in the administrative area of the associated MSC. A VLR can be responsible for areas of more than one MSC. A mobile station (MS) roaming freely may be registered in a VLR of its home network or the Foreign network depending upon its current location.

Channels in GSM

The logical channels are divided into two categories:

Traffic channels (TCH) and Signaling channels (BCH).

Traffic channels (TCH)

The traffic channel is used for the transmission of the user payload, which could be data or speech. They do not carry any control information.

Broadcast channels (BCCH)

This is a unidirectional point to multipoint channel between the base station and the mobile station. The following kinds of information are sent on this channel, radio channel configuration is about the current cell nearby cells, synchronization information reduces its frequency and frame numbering, registration information LAI, CI, BSIC.

Modulation for GSM

GSM uses Gaussian Minimum shift keying (GMSK) for modulation. It belongs to the family of continuous phase modulation schemes.

Multiple access for GSM

- GSM uses a combination of FDMA and TDMA. The two frequency bands 45 MHz apart have been reserved for GSM: one is the 890–915 MHz for uplink and 935–960 MHz for downlink.

- Each of these bands which are of 25 MHz are then further subdivided into 124 single carrier channels of 200 kHz. There exists a guard band of 200 kHz in each uplink or downlink band.
- Now each of the two hundred kilohertz channels carries a 8-time division multiple access channel, that is, 8 time slots.
- Now the TDMA frames of the uplink are transmitted with a delay of 3 time slots. The mobile station uses the same time slot in uplink and downlink since it does not send and receive in the same time slot.
- Each TDMA frame lasts 156.25 bit times and, when used, contains a data burst. The time slot is 576.9 microseconds and the frame length is 4.613 milliseconds; this is the TDMA part of the GSM network.

GSM Services

GSM offers following services:

- Telephony.
- Facsimile group 3 (E1).
- Emergency calls.
- Short message service.
- Fax mail.
- Voice mail.

Disadvantage of GSM

- Each radio channel uses a frequency guard band, which is inefficient.
- Complex frequency planning is required for avoiding a co-channel and an adjacent channel interference.
- It uses a combination of FDMA and TDMA. A time slot has to be given as a time guard. So, every time a slot requires a guard band before and after so as not to emerge into the next adjacent time slot, making it inefficient. A time slot is occupied even when there is a pause in the speech.

8.4 Cellular Concepts

- A cellular network is a radio network distributed over land areas called cells, each served by at least one fixed-location transceiver known as a cell site or base station.
- These cells when joined provide radio coverage over a wide geographic area.

- This enables a large number of portable transceivers to communicate with each other and, with fixed transceivers and telephones anywhere in the network, via base stations, even if some of the transceivers move through more than one cell during transmission.

8.4.1 Basic Terminology in Cellular Communication

Mobile: It is a radio terminal attached to a high-speed mobile platform. (e.g. a cell phone in a fast moving vehicle)

Subscriber: A mobile or a portable user.

Base stations: Base stations are fixed antenna units with which the subscribers communicate. Base stations are connected to a commercial power source and a backbone network.

Cells: The area of coverage is usually divided into cells. Each cell has its own base station usually located at its center or sometimes at the edge.

Control channels: These are radio channels used for call set up, call request, and call initiation.

Forward channel (downlink): It represents the radio channel used for transmission of information from the base station to the mobile station.

Reverse channel (uplink): The uplink or the reverse channel is the radio channel used for transmission of information from the mobile to the base station.

Foot print: The actual coverage area of a cell is called as foot print.

Cellular system: The cellular system is shown in Figure 8.4. The basic components of cellular systems are:

Mobile station (MS): Mobile handsets which are used by user for communication with other users.

Base Stations (BS): Each cell contains an antenna, which is controlled by a small office.

Mobile Switching Center (MSC): Each base station is controlled by a switching office, called mobile switching center.

MSC Databases

- Home location register database (HLR)—stores information about each subscriber belonging to it
- Visitor location register database (VLR)—maintains information about subscribers roaming within a MSC location area
- Authentication center database—used for authentication of activities, holds encryption keys

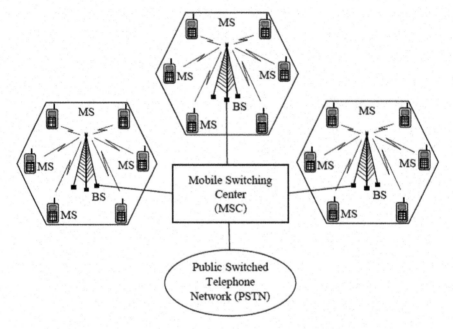

Figure 8.4 Cellular system.

- Equipment identity register database (EIR)—keeps track of the type of equipment that exists at the mobile station

Public switched telephone network (PSTN): The public switched telephone network (PSTN) is the network of the world's public circuit switched telephone networks.

Connection between MS, BS, MSC, and PSTN

The connection between MS, BS, MSC, and PSTN is shown in Figure 8.5.

- The service coverage area of a cellular network is divided into many smaller areas, referred to as cells, each of which is served by a base station (BS).
- The BS is fixed, and connected to the mobile telephone switching office (MTSO), also known as the mobile switching center (MSC).
- An MTSO is in charge of a cluster of BSs and it is, in turn, connected to the PSTN.
- MSs such as cell phones are able to communicate with wire line phones in the PSTN.
- Both BSs and MSs are equipped with a transceiver.

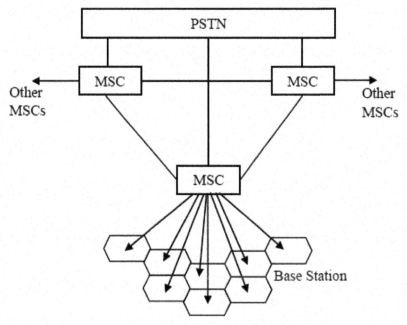

Figure 8.5 Connection between BS, MS, MSC, and PSTN.

Traffic and Control channels
Two types of channels are available between the mobile station and the base station

- Control channels: used for exchange of information having to do with setting up and maintaining calls
- Traffic channels: carry voice or data connection between users

Both voice and control channels are further divided into forward (or downlink) and reverse (or uplink).

- A *forward channel* is used for carrying traffic from the *BS to the MS*.
- A *reverse channel* is used for carrying traffic from the *MS to the BS*.

Various process involved in call initiation
At transmitting end:

1. A caller enters a 10-digit code (phone number) and presses the call button.
2. The MS scans the band to select a free channel and sends a strong signal to send the number entered to BS.
3. The BS relays the number to the MSC.

4. The MSC in turn dispatches the request to all the base stations in the cellular system.
5. The Mobile Identification Number (MIN) is then broadcast over all the forward control channels throughout the cellular system. This is known as paging.
6. The MS responds by identifying itself over the reverse control channel.
7. The BS relays the acknowledgement sent by the mobile and informs the MSC about the handshake.
8. The MSC assigns an unused voice channel to the call and call is established.

At receiving end

1. All the idle mobile stations continuously listen to the paging signal to detect messages directed at them.
2. When a call is placed to a mobile station, a packet is sent to the receiver's home MSC to find out where it is.
3. A packet is sent to the base station in its current cell, which then sends a broadcast on the paging channel.
4. The receiver MS responds on the control channel.
5. In response, a voice channel is assigned and ringing starts at the MS.

8.5 Frequency Reuse and Handoff

Need for frequency reuse

- Suppose there are fixed telephone networks and they were running wires to every household. Every household would be given its own allocation of radio spectrum for an analog speech of 4 kHz bandwidth. The number of households in Chennai may be assumed as about 12.5 million.
- Hence, the need is 12.5 million × 4 kHz = 50 GHz. Clear allocation of this volume of spectrum is not possible. Therefore, the spectrum needs efficient use.

Concept of frequency reuse

Cellular systems rely on an intelligent allocation and reuse of channels throughout the coverage area. Each base station is allocated a group of radio channels for use within a small geographic area of its cell. Base stations in neighboring cells are assigned completely different sets of channel frequencies.

The cellular capacity or to accommodate more number of users, the same set of channels may be used for increase and covering different cells separated from one another by a distance large enough to keep interference level within tolerable limits. Figure 8.6 shows three groups of cells called *clusters*. Each cluster has seven cells, cells with the same letter using the same set of frequencies. The design procedure for allocating channel groups for all the cellular BS within the system is called *frequency reuse or frequency planning*. In this figure, letters *A, B, C, D, E, F,* and *G* denote seven different frequencies.

An interference coming from a cell and using the same frequency is called a ***co-channel interference***.

A large reuse distance has to be used for avoiding co-channel interference. The closest distance between the centers of two cells using the same frequency is determined by *reuse distance*.

It is given by

$$D = R\sqrt{3N}$$

where *R*-cell radius

N-Number of cells per cluster

D-reuse distance

Reuse factor: A fraction of the total available channels assigned to each cell within a cluster is 1/N, which is called a *reuse factor*.

It is given by,

$$q = \frac{D}{R} = \sqrt{3N}$$

When the total of *M* channels is allocated for cellular communications and when the coverage area consists of *N* cells, there are *MN/7* channels available

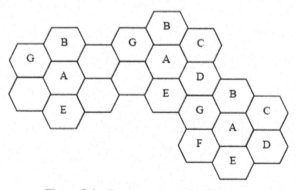

Figure 8.6 Frequency reuse-7-cell pattern.

in the coverage area for concurrent use based on the seven-cell reuse pattern. The Figure 8.7 shows the reuse factor of 1/4 and 1/7.

Techniques for increasing the capacity of cellular networks

1. Adding new channels.
2. Frequency borrowing—frequencies are taken from adjacent cells by congested cells.
3. Cell splitting—cells in areas of high usage can be split into smaller cells. Figure 8.8 shows the structure of cell splitting.
4. Cell sectoring—cells are divided into a number of wedge-shaped sectors, each with its own set of Channels.
5. Microcells—antennas move to buildings, hills, and lamp posts.

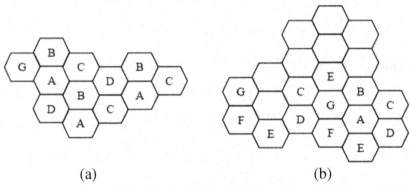

(a)   (b)

Figure 8.7   (a) Reuse factor 1/4. (b) Reuse factor 1/7.

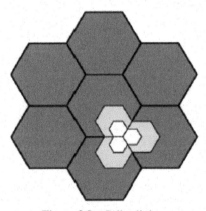

Figure 8.8   Cell splitting.

8.5.1 Handoff

It is defined as follows: when a mobile moves into a different cell while the call is in progress, the mobile switching center (MSC) automatically transfers the call to a new channel belonging to the new base station (BS).

- The handoff operation involves identifying a new BS and the allocation of voice and control signals associated with the new BS.
- Handoff must be performed successfully, as infrequently as possible, and must be imperceptible to the user.

Handoff region

Figure 8.9 shows the variation of signal strength from either base station. It is possible to decide on the optimum area in which the handoff can take place. Signal strength is due to $BS_{OLD} = P_i(x)$ and signal strength is due to $BS_{NEW} = P_j(x)$. By looking at the variation of signal strengths from either base station, it is possible to decide on the optimum area where handoff can take place with minimum receiver sensitivity.

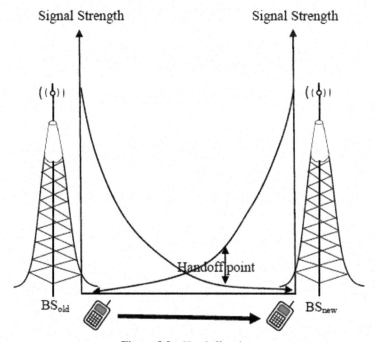

Figure 8.9 Handoff region.

Handoff is made when the received signal at the BS falls below a pre-specified threshold. In deciding when to hand off, it is important to ensure that the drop in the signal level is not due to momentary fading. The BS monitors the signal for a certain period of time before initiating handoff. The length of time needed for deciding the necessity of the handoff depends on the speed at which the mobile is moving.

Figure 8.10 shows two situations: One where the handoff occurs successfully in the lower curve and the second when it fails. So the lower curve has first to be looked into. There are two regions, namely base stations 1 and 2 and user moving from left to right. Here is a depiction of the falling of the signal strength of the first base station. But very soon it locks on to another signal strength of base station B. So there is a sudden jump in the power signal level because the user gets a much stronger signal from base station B. So the user is hooked on to the next power level. Therefore, it is a successful handoff. On the other hand (the upper part of the figure), keeps on dropping and is unable to latch onto nearer base station. It is due to the absence of availability of any channel or the presence of some other problems.

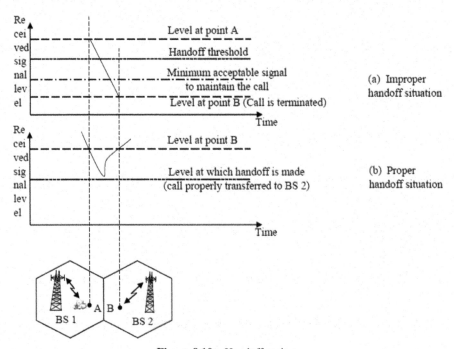

Figure 8.10   Handoff region.

Handoff strategies

- In the first-generation analog cellular systems, the signal strength is measured by the base stations and supervised by the mobile switching center.
- In the second generation, those which use TDMA technology mobile-assisted handoffs are called MAHO. Wherein, every mobile station measures the received power to the surrounding base stations and continually reports these values to the corresponding base stations.
- Handoff is initiated when the signal strength of a neighboring base station exceeds that of the current base stations. So attempt is made to hook on to a stronger signal at any point of time.

Soft handoff

The ability to make a choice between the instantaneous received signals from different base stations is called *Soft handoff*. CDMA uses soft handoff approach.

8.6 Multiple Access Schemes

Simultaneous access to communication resources for many subscriber from widely different locations on earth is always preferred. Such a capability is called "Multiple Access".

The purpose of multiple access is to permit sharing of the communication resources by a large number of subscribers (users) seeking to communicate with each other. It is also desired that the sharing of the communication resources should be accomplished without causing serious interference to each other.

There are three basic types of multiple access techniques. They are

1. Frequency Division Multiple Access (FDMA)
2. Time Division Multiple Access (TDMA)
3. Code Division Multiple Access (CDMA).

8.6.1 FDMA

In this FDMA technique, disjoint frequency sub-bands are allocated to different users on a continuous time basis. In TDMA, guard bands are introduced as buffer zones between the allocated adjacent channel bands as shown in the Figure 8.11. This is done for avoiding interference between users.

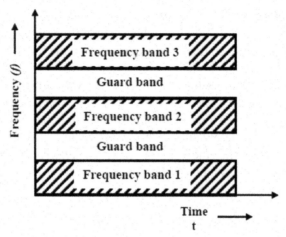

Figure 8.11   Frequency division multiple access.

In FDMA, the user should occupy the available time slot fully, but in a small different frequency slot.

Channels can be assigned to the same frequency at all times (pure FDMA). Sometimes frequencies are changed according to a certain pattern (FDMA combined with TDMA). Frequency Division Multiplexing (FDM) is often used for simultaneous access to the medium by a base station and a mobile station in cellular networks. Here, the two partners typically establish a duplex channel. The signals from the base station to the mobile station and vice versa are separated using different frequencies. This method is called as Frequency Division Duplex (FDD). Both the partners have to know the frequencies in advance or otherwise they cannot listen into the medium. The two frequencies are also known as uplink (i.e.) from mobile station (or mobile unit) to base station and as downlink (i.e.) from base station to mobile station.

For explaining FDMA, let us consider a mobile phone network based on the GSM standard for 900 MHz. The basic frequency allocation scheme for GSM is fixed. All uplinks use the band between 890 and 915 MHz and all downlinks use the band between 935 and 960 MHz. Both the uplink and downlink have been allocated 25 MHz for a total of 50 MHz. The total number of channels in FDMA is 124 and each channel is 200 kHz. The duplex separation is 45 MHz.

According to FDMA, the base station allocates a certain frequency for up- and downlink to establish a duplex channel with the mobile unit.

Up- and downlinks have a fixed relation. If the uplink frequency is
$f_u = 890\,\text{MHz} + (n \times 0.2\,\text{MHz})$,

the downlink frequency is $f_d = f_u + 45\,\text{MHz}$.

i.e. $f_d = 890\,\text{MHz} + (n \times 0.2\,\text{MHz}) + 45\,\text{MHz} = 935\,\text{MHz} + (n \times 0.2\,\text{MHz})$ for a certain channel n.

This shows the use of FDM for multiple access and duplex according to a predetermined scheme (Figure 8.12).

8.6.2 TDMA

In the TDMA technique, each user is allocated the full spectral occupancy but only for a short duration of time called a time slot. Buffer zones in the form of guard band are inserted between the assigned time slots as shown below to avoid interference between users.

An alternative to FDMA is the TDMA technique. In TDMA, each user has access to the entire authorized radio frequency spectrum for a short time for transmitting messages. Every user shares the authorized frequency spectrum with all other users who have time slot allocations at other pre-assigned times (Figure 8.13).

Depending upon the allocation of available radio spectrum, the TDMA technique is divided into two types. They are

1. Narrow-band TDMA
2. Wide-band TDMA

When available frequency spectrum is only partially allocated to a particular user group, the access method is called as Narrow-band TDMA.

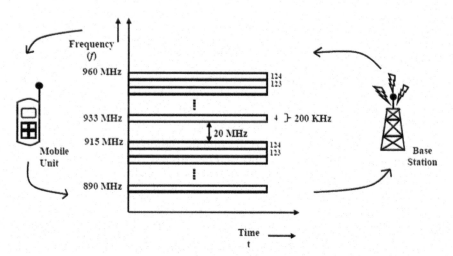

Figure 8.12 FDD/FDMA system.

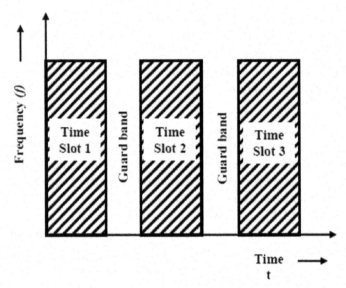

Figure 8.13 Time division multiple access.

When the entire available frequency spectrum is allocated to each user during the users' time slot, this access method is called as Wide-band TDMA.

It is important here to note that listening to different frequencies at the same time is quite difficult but listening to many channels separated in time at the same frequency is simple. This is the basic concept behind the TDMA technique.

In the TDMA technique, TDM patterns are used to implement multiple access and a duplex channel between a base station and mobile station as shown in the figure below. The TDMA technique is used in DECT cordless phone system. Allocation of different time slots for uplink and downlink using the same frequency is called Time Division Duplex (TDD).

The figure shows the time division multiplexing for multiple access and duplex operation. As shown in Figure 8.14, the base station uses one out of 12 time slots for downlinks, whereas the mobile station uses one out of 12 time slots for uplink. Hence, the uplink and downlink systems are separated by time slot. In addition, up to 12 different mobile stations can use the same frequency without causing interference to others. In the above example (i.e. DECT cordless phone system), the access pattern is repeated every 10 ms (i.e., till each time a slot has a duration of 417 µs). This repetition ensures access to the medium every 10 ms. Thus, TDM is allocating time slots for channels in

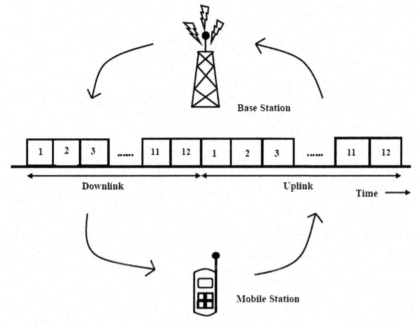

Figure 8.14 TDD/TDMA scheme.

a fixed pattern, which results in a fixed bandwidth and is the typical solution for any wireless phone system.

In this TDMA technique, there is need for synchronization in the time domain between the sender and the receiver. This synchronization can be done by using a fixed pattern (i.e., allocating a certain time slot for a channel) or by using a dynamic allocation scheme.

8.6.3 CDMA

Codes with certain characteristics can be applied to the transmission to enable the use of Code Division Multiplexing (CDM). CDMA uses the codes to separate different users in code space and to enable access to a shared medium without interference.

CDMA technique is a hybrid combination of FDMA and TDMA as illustrated in the figure.

Frequency hopping
In this technique, during each successive time slot, the frequency bands assigned to the users are recorded in an essentially random manner.

For example, during time slot 1, user 1 occupies frequency band 1, user 2 occupies frequency band 2, user 3 occupies frequency band 3, and so on. During time slot 2, user 1 hops to frequency band 3, user 2 hops to frequency band 1, user 3 hops to frequency band 2, and so on. Such arrangement is called "frequency hopping" (Figure 8.15).

In the CDMA technique, frequency hopping mechanism can be implemented through the use of a pseudo-noise (PN) sequence, which is a cyclic code with noise-like characteristics.

An important advantage of CDMA over both FDMA and TDMA is that it can provide for "secure communications".

In CDMA, each user is provided with an individual distinctive pseudo-noise (PN) code. In the absence of mutual correlation, within the same mobile cell, a large number of independent users can transmit at the same time and in the same radio bandwidth. The receiver decorrelates (despread) the information and regenerates only the desired data sequence.

Here, each mobile unit transmits its data source along with the uncorrelated pseudo-noise code in the same radio frequency bandwidth at the same time as the other mobile transmitters. The base station receiver receives all the signals and the adaptive power circuitry at the base station ensures that all the received signals at the base station have the same power. The base station receiver decorrelate (despread) and demodulate the independent messages from the mobile transmitter.

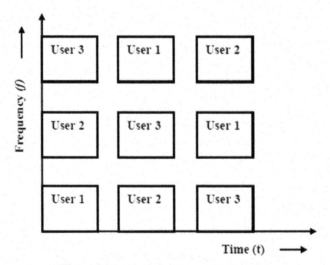

Figure 8.15 Code division multiple access.

The main problem arising in CDMA is how to find 'good' codes and how to separate the signal from the noise generated by other signals and the environment.

The following theoretical example explains the basic function of CDMA (Figure 8.16).

Two senders A and B want to send data CDMA and assign the following key sequence key $A_K = 010011$, key $B_K = 110101$. Let us assume that sender A sends the bit $A_d = 1$ and B sends the data bit $B_d = 0$. For illustration, let us code a binary 0 as –1 and as a binary 1 as +1.

Both the senders spread their signal using their keys as chipping sequence. The term spreading refers to the simple multiplication of the data bit with the whole chipping sequence.

Sender A then sends the signal A_s as,
$$A_s = A_d \times A_k = (+1) \times (-1, +1, -1, -1, +1, +1)$$
$$= (-1, +1, -1, -1, +1, +1)$$

Sender B then sends the signal
$$B_s = B_d \times B_k = (-1) \times (+1, +1, -1, +1, -1, +1)$$
$$= (-1, -1, +1, -1, +1, -1)$$

Both the signals are then transmitted simultaneously using the same frequency and thus the signals get superimposed in space.

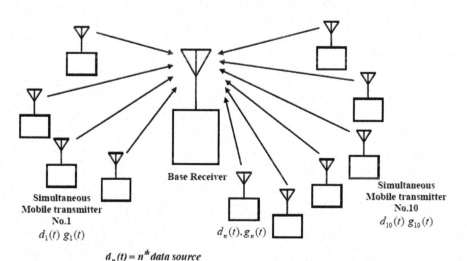

$d_n(t) = n^{th}$ *data source*

$g_n(t) = n^{th}$ *pseudonoise code*

Figure 8.16 CDMA access mechanisms.

(i.e.,) Assume any error occurring from environmental noise.

Neglecting interference from other senders and environmental noise and assuming that the signals have the same strength at the receiver, the signal received at the receiver is given by

$C = A_g + B_g = (-2, 0, 0, -2, +2, 0)$

The receiver now wants to receive data from sender A and therefore tunes into the code of A (i.e.,), and it applies A's code for despreading as

$$C \times A_k = (-2, 0, 0, -2, +2, 0) \times (-1, +1, -1, -1, +1, +1)$$
$$= (2 + 0 + 0 + 2 + 2 + 0)$$
$$= 6$$

As the result is much larger than 0, the receiver detects as binary 1.

If the receiver wants to receive data from sender B, then it applies B's code for despreading as

$$C \times B_k = (-2, 0, 0, -2, +2, 0) \times (+1, +1, -1, +1, -1, +1)$$
$$= (-2 + 0 + 0 - 2 - 2 + 0)$$
$$= -6$$

As the result is negative, the receiver detects as binary 0.

In this case, assume that noise and interference are introduced, and it is difficult for the receiver to detect the original data bits.

8.6.4 Comparison of FDMA, TDMA, and CDMA

Sl. No.	Approach	FDMA	TDMA	CDMA
1.	Idea	Segment the frequency band into disjoint sub-bands	Segment the sending time into disjoint time slots, either fixed or demand-driven patterns	Spread the spectrum using orthogonal codes
2.	Terminals	Every terminal has its own frequency, uninterrupted	All terminals are active for short periods of time on the same frequency	All terminals can be active at the same place at the same moment, uninterrupted
3.	Signal Seperation	Done by means of filtering in the frequency domain	Done by means of synchronization in the time domain	Done by means of using codes and special receiver

(Continued)

4.	Advantages	Simple and robust	Fully digital and very flexible	Flexible, less planning needed, and soft handoff
5.	Disadvantages	Inflexible frequencies are a scare (rare) resource	Need for the guard space and the synchronization is difficult	Receiver has higher complexity, needs more complicated power control for the senders
6.	Comment	Typically combined with TDMA and SDMA	Standard in fixed mobile networks, together with FDMA and SDMA	Higher complexity and this will be integrated with FDMA/TDMA

Table Continued

8.7 Satellite Communication

- Satellite is a powerful long-distance and point-to-multi point communication system. In other words, communication satellite is a radio frequency (RF) repeater.
- To overcome the limitations of cellular communication, which covers only a 2 km region, the transmitting antenna is placed on the satellite and the satellite is placed in the orbit high above the earth.
- The function of the satellite is to communicate between different earth stations around the earth. Hence, with the help of satellite, it is easy to communicate over thousands of kilometer.

For example, the entire earth can be covered with the help of three numbers of geo-synchronous satellites.

Elements of satellite communication

The major elements of satellite communication systems are:
  Ground segment and space segment.

- The ground segment consists of earth stations and network control centers of the entire satellite system.
- Space segment consists of the spacecraft and launch mechanism.

Figure 8.17 shows the various segments of the satellite system. The communication from ground station (or earth station) to the satellite is referred to as *uplink transmission* and communication from the satellite to earth station is called *downlink transmission*. The ***major blocks of satellite systems are***

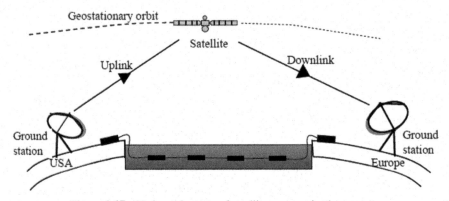

Figure 8.17   Various elements of satellite communication system.

1. uplink
2. Satellite transponder and
3. downlink system.

8.7.1 Satellite Uplink System Model

Figure 8.18 shows a detailed structure of the uplink system. The major parts of the uplink system include modulator, intermediate frequency (IF) section, mixer or IF to RF conversion block, high-power amplifier, and transmitting antenna.

- The inputs from various sources are frequency multiplexed by FDM (frequency division multiplexing) and this input signals are of low frequency (base band) in nature.
- Digital modulation techniques are performed at transmitter side for conversion to band pass signals. The various types of digital modulation techniques are PSK, QPSK, and QAM.

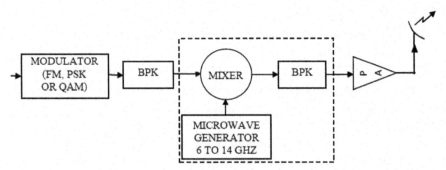

Figure 8.18   Satellite uplink system model.

- The signals are filtered by band pass filter for eliminating noises in band pass signals.
- *Frequency conversion:* Frequency of operation of satellite systems is either in C band (uplink frequency of 6 GHz and downlink frequency of 4 GHz) or in Ku band (uplink frequency of 14 GHz and downlink frequency of 12 GHz).
- The inputs to the mixer circuits are band pass in nature and the frequencies are in the range of MHz only. Hence, mixer circuits are used for conversion of MHz to GHz range.
- Mixer circuits are electronic circuits. They accept two inputs: the inputs are band pass input signal $= f_1$ (in MHz range) and locally generated RF signals $= f_2$ (6 GHz or 14 GHz). They produce two outputs, sum of outputs $(f_1 + f_2)$ and difference output $(f_1 - f_2)$.
- Then $(f_1 + f_2)$ output is selected by suitable filter circuits, since RF signal is needed in the GHz range. The transmitting signals are now ready with a proper frequency.
- RF signals are amplified by a high-power amplifier (HPA). This HPA provides adequate input sensitivity and output power for propagation of the signal to the satellite transponder. The commonly used high-power amplifiers are travelling wave tube (TWT) amplifiers. Finally, amplified signals propagate to satellite transponder with the help of transmitting antenna.

8.7.2 Satellite Transponder

- It consists of a frequency conversion block, a low noise amplifier, and a band pass filter. The circuit diagram is shown in Figure 8.19.
- The satellite transponder receives the signal from the earth station. A satellite system is assumed to operate in C band. Hence, received

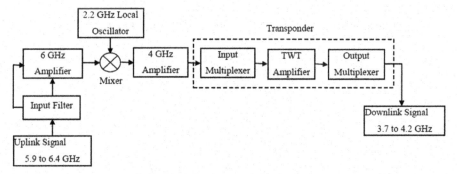

Figure 8.19 Satellite transponder.

signals are in the frequency range of 6 GHz. The received signals are amplified by an amplifier. These amplified signals are given to a mixer (or frequency conversion) circuit.

- *Frequency conversion:* Downlink frequencies, it is well known, are 4 GHz only. Therefore, down conversion of the frequency from 6 to 4 GHz is needed. This can be done by mixer circuits, which are electronic circuits. They accept two inputs: the first input is a 6 GHz received signal (f_1) and the second is a locally generated 2 GHz signal (f_2). Mixer circuits produces two outputs in the form of $(f_{1'} + f_2)$ and $(f_{1'} - f_2)$.
- $(f_{1'} - f_2) = 4$ GHz signal can be selected by using a filter circuit.
- The low-level-power amplifier, which is commonly a travelling wave tube (TWT)M, amplifies the RF signal for transmission through the downlink earth stations.

8.7.3 Satellite Downlink System Model

Figure 8.20 illustrates the detailed structure of a downlink system model. The major parts of the downlink system include demodulator, intermediate frequency (IF) section, mixer or RF-to-IF conversion block, band pass filters, and receiving antenna.

- In the downlink system model, downlink earth stations receive the signals from satellite transponder. Received signals are in the frequency of 4 GHz and it is amplified by a low-noise amplifier (LNA).
- *Frequency conversion:* Mixer circuits are used for down conversion of the received RF signal to IF signal.

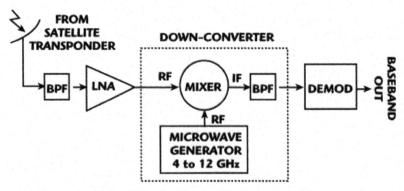

Figure 8.20 Satellite downlink system model.

- Received signals at the IF stage are in MHz range. Performance of a suitable demodulation operation at earth station is necessary for conversion into a base band signal.

Different orbits of satellite systems are
1. LEO-Lower earth orbit-Satellites are located at 500–1500 km from earth
2. MEO-Medium earth orbit-satellites are located at 6000–2000 km from earth
3. GEO-Geo-synchronous orbit –satellites are located at 36,000 km from earth surface.

8.8 Bluetooth

- Bluetooth is a radio frequency specification for short range, point-to-point, and point-to-multi-point voice and data transfer.

In spite of facilitating the replacement of cables, Bluetooth technology works as a universal medium to bridge the existing data networks, a peripheral interface for existing devices and provides a mechanism to form a short ad hoc network of connected devices away from fixed network infrastructures.

The technology of Bluetooth centers around a 9×9 mm microchip, which functions as a low-cost and short-range radio link. The technology provides a 10 m personal bubble that supports simultaneous transmission of both voice and data for multiple devices. Devices up to eight can be connected in a piconet, and up to 10 piconets can exist within the 10 m bubble. Each piconet can support up to 3 simultaneous full duplex voice devices.

Bluetooth wireless technology is designed to be as secure as a wire with public/private key authentication up to 128-bit, and streaming cipher up to 64 bits based on A5 security.

Transmission type and rate
The baseband (single channel per line) protocol combines circuit and packet switching. Packets do not arrive out of order, slots (up to five) can be reserved for synchronous packets for ensuring this. As noted earlier, a different hop signal is used for each packet. Circuit switching can be either asynchronous or synchronous. Up to three synchronous (voice) data channels, or one synchronous and one asynchronous data channel, can be supported on one channel. Each synchronous channel can support a 64-Kb/s transfer rate, which is quite adequate for voice transmission. An asynchronous channel

can transmit as much as 721 Kb/s in one direction and 57.6 Kb/s in the opposite direction. It is also possible for an asynchronous connection to support 432.6 Kb/s in both directions when the link is symmetric.

Figure 8.21 shows the network arrangement of Bluetooth technology:

Security features
Security can be provided in three ways. They are Pseudo-random frequency band hops, authentication, and encryption.

Bluetooth connections are established by following techniques
1. Standby
2. Page/Inquiry
3. Active
4. Hold
5. Sniff
6. Park

Bluetooth model is used for (i) Voice and data points, (ii) Peripheral interconnects, and (iii) Personal area networking (PAN)

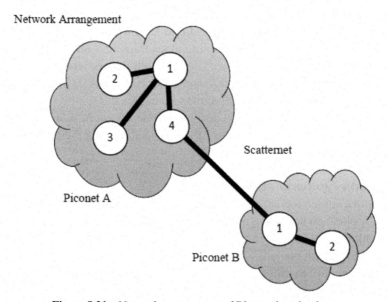

Figure 8.21   Network arrangement of Bluetooth technology.

- Voice/data access points constitute one of the key initial usage models and involve connecting a computing device to a communicating device via a secure wireless link.
- The second category of uses, *peripheral interconnects*, involves connecting other devices.
- The last usage model, *Personal Area Networking (PAN)*, focuses on the ad-hoc formation and breakdown of personal networks.

Bluetooth characteristics

- It separates the frequency band into hops. This spread spectrum is used for hopping from one channel to another, which process adds a strong layer of security.
- Up to eight devices can be networked in a piconet.
- Signals can be transmitted through walls and briefcases, thus eliminating the need for line-of-sight.
- Devices do not need to be pointing at each other, as signals are omni-directional.
- Governments worldwide regulate it, so it is possible to utilize the same standard wherever one travels.

Advantages of Bluetooth

- No line of site restrictions as with IrDA.
- Power consumption gets integrated in battery-powered devices very practical.
- 2.4-GHz radio frequency ensures worldwide operability.
- Tremendous momentum not only within the computer industry but also in other industries like cellular telephones and transportation.

Applications
1. The Internet Bridge
An extension of this model could be a mobile computer that allows surfing the Internet irrespective of the location of the user, and regardless of whether the user is cordlessly connected through a mobile phone (cellular) or through a wire line connection (e.g. PSTN, ISDN, LAN, xDSL).
2. Automatic Check-in
Hotels are testing, or plan to test, services that allow guests to check in, unlock room doors and even control room temperature with handheld devices equipped with Bluetooth Technology.

References

[1] H. Taub and D.L. Schilling. *Principles of communication*. McGraw-Hill, 2003.

[2] S. Haykin. *Digital communications*, John Wiley & Sons, 2001.

[3] K. Sam shanmugam. *Digital and analog communication systems*, John Wiley & Sons, 2005.

[4] S. Haykin. *An introduction to analog and digital communications*, John Wiley & Sons, 2004.

[5] J.G. Proakis. *Digital communications*. McGraw-Hill, 2001.

[6] B.P. Lathi. *Modern digital and analog communication systems*. Oxford University Press, 2003.

[7] B. Sklar and P. Kumarray. *Digital communications-Fundamentals and applications*. Pearson Education, 2001.

[8] R.E. Ziemer and W.H. Tranter. *Principles of* communications. Jaico Book Publishing, 2001.

[9] S. Haykin. *Communication systems*, John Wiley Edition Publishing Company, 2009.

[10] B.A. Forouzan. *Data communication and networking*. McGraw-Hill, 2003.

[11] J. Schiller. *Mobile communications*. Pearson Education, 2003.

[12] T.K. Moon. *Error correction coding-Mathematical methods and algorithms*. Wiley Interscience, 2006.

[13] T.T. Ha. *Digital satellite communications*. McGraw-Hill, 2014.

[14] K. Feher. *Wireless digital communications-Modulation and spread spectrum applications*. Eastern Economy Edition, 2002.

[15] R. Blake. *Wireless communication technology*. Delmar Publishing Company, 2001.

237

[16] A.F. Molisch. *Wireless communication*. John Wiley Publishing Company, 2011.

[17] W.C.Y. Lee. *Mobile cellular telecommunications-Analog and digital systems*. McGraw-Hill Edition, 2006.

[18] T.S. Rappaport. *Wireless communications-Principles and practice*. Pearson Education, 2003.

Index

About the Authors

M. A. Bhagyaveni graduated from Government college of Technology, Coimbatore, received the M.E. degree in the field of optical communication from Anna University. She has obtained her Ph.D. in the field of Adhoc Networks. She is presently working as Professor in the Department of Electronics and Communication Engineering, College of Engineering, Anna University, Chennai. Her field of interest includes Wireless Communication, Digital Communication, Sensor Networks and Cloud Computing. She has completed one project for CVRDE, Avadi and pursuing one DST project. She has published many research paper in National/International Journals and Conferences.

R. Kalidoss completed bachelor degree in Electronics and Communication Engineering from Madurai Kamaraj University and master degree in Communication Systems from Anna University, Chennai. Further, he obtained doctoral degree from Anna University, Chennai in the field of Cognitive Radio. He is currently working as an Associate Professor in Department of Electronics and Communication Engineering in SSN College of Engineering, Chennai. His current research interests include Adaptive Channel Modelling in Cognitive Radio, Advanced Spectrum Utilization and Cognitive Radio architecture. He has published many research articles in refereed International/National Journals/Conferences.

K. S. Vishvaksenan received the MS degree from Birla institute of Technology and Science (BITS, Pilani), Rajasthan, India. He obtained his M.E degree, Electronics and communication Engineering from College of Engineering, Guindy, Anna University, Chennai. He has been awarded PhD degree from College of Engineering, Guindy, Anna University, Chennai, India in the year 2013. His research field includes MIMO, channel encoder, multiuser communications, transmitter pre-processing and Cognitive Radio based WRAN.

Lightning Source UK Ltd.
Milton Keynes UK
UKOW06n1147270117
293027UK00002B/29/P